John Timbs

Eccentricities of the animal creation

John Timbs
Eccentricities of the animal creation
ISBN/EAN: 9783337228620
Printed in Europe, USA, Canada, Australia, Japan
Cover: Foto ©berggeist007 / pixelio.de

More available books at **www.hansebooks.com**

ECCENTRICITIES

OF

THE ANIMAL CREATION.

BY JOHN TIMBS,

AUTHOR OF "THINGS NOT GENERALLY KNOWN."

WITH EIGHT ENGRAVINGS.

SEELEY, JACKSON, AND HALLIDAY, 54, FLEET-STREET.
LONDON. MDCCCLXIX.

CONTENTS.

INTRODUCTORY.—CURIOSITIES OF ZOOLOGY.

Natural History in Scripture, and Egyptian Records, 11.—Origin of Zoological Gardens, 12.—The Greeks and Romans, 12.—Montezuma's Zoological Gardens, 13.—Menagerie in the Tower of London, 14.—Menagerie in St. James's Park, 14.—John Evelyn's Notes, 15.—Ornithological Society, 15.—Continental Gardens, 16.—Zoological Society of London instituted, 16; its most remarkable Animals, 16.—Cost of Wild Animals, 18.—Sale of Animals, 20.—Surrey Zoological Gardens, 20.—Wild-beast Shows, 21.

THE RHINOCEROS IN ENGLAND.

Ancient History, 22, 23.—One-horned and Two-horned, 25, 26.—Tractability, 25.—Bruce and Sparmann, 27.—African Rhinoceros in 1868, 27.—Description of, 29.—Burchell's Rhinoceros, 30.—Horn of the Rhinoceros, 31, 32.

STORIES OF MERMAIDS.

Sirens of the Ancients, 33.—Classic Pictures of Mermaids, 34.—Leyden's Ballad, 35.—Ancient Evidence, 36, 37, 38.—Mermaid in the West Indies, 39.—Mermaids, Seals, and Dugongs, 41.—Mermaids and Manatee, 42.—Test for a Mermaid, 43.—Mer-

A 2

maid of 1822, 43.—Japanese Mermaids, 44.—Recent Evidence, 47, 48.

IS THE UNICORN FABULOUS?

Ctesias and Wild Asses, 65.—Aristotle, Herodotus, and Pliny, 50.—Modern Unicorns, 50.—Ancient Evidence, 51.—Hunting the Unicorn, 52.—Antelopes, 53, 54.—Cuvier and the Oryx, 54.—Tibetan Animal, 55.—Klaproth's Evidence, 55.—Rev. John Campbell's Evidence, 57.—Baikie on, 58.—Factitious Horns in Museums, 59.—Unicorn in the Royal Arms, 60.—Catching the Unicorn, 60.—Belief in Unicorns, 61.

THE MOLE AT HOME.

Economy of the Mole, 62.—Its Structure, 63.—Fairy Rings; Feeling of the Mole, 64.—Le Court's Experiments, 62, 65.—Hunting-grounds, 67.—Loves of the Moles, 68, 69.—Persecution of Moles.—Shrew Mole, 70.—Hogg, the Ettrick Shepherd, on Moles, 71.

THE GREAT ANT-BEAR.

The Ant-Bear of 1853, 72, 73.—Mr. Wallace, on the Amazon, describes the Ant-Bear, 73.—Food of the Ant-Bear, 74.—His Resorts, 75.—Habits in Captivity, by Professor Owen, 76 80.—Fossil Ant-Bear, 80, 81.—Tamandua Ant-Bear, 82—Von Sack's Ant-Bear, 83.—Porcupine Ant-Eater, 84.—Ant-Bears i the Zoological Gardens, 84.

CURIOSITIES OF BATS.

Virgil's Harpies, 85.—Pliny on the Bat, 85.—Reremouse and Flittermouse, 86.—Bats, not Birds but Quadrupeds, 87.—Sir Charles Bell on the Wing of the Bat, 87.—Vampire Bat from Sumatra, 88.—Lord Byron and Vampire, 89.—Levant Superstition, 89.—Bat described by Heber, Waterton, and Steadman, 90.—Lesson on Bats, 91.—Bat Fowling or Folding, 91, 92.—

Sowerby's Long-eared Bat, 92, 96.—Wing of the Bat, 96.—*Nycteris* Bat, 97.—*Kalong* Bat of Java, 98.—Bats, various, 100, 101.

THE HEDGEHOG.

Hedgehog Described, 102.—Habits, 103.—Eating Snakes, 105.—Poisons, 105, 106.—Battle with a Viper, 105.—Economy of the Hedgehog, 106, 107.

THE HIPPOPOTAMUS IN ENGLAND.

Living Hippopotamus brought to England in 1850, 108.—Capture and Conveyance, 111.—Professor Owen's Account, 111-115.—Described by Naturalists and Travellers, 115-118.—Utility to Man, 118-119.—Ancient History, 119.—In Scripture, 120.—Alleged Disappearance, 121.—Fossil, 122.

LION-TALK.

Character, 123.—Reputed Generosity, 125.—Burchell's Account, 125.—Lion-Tree in the Mantatee Country, 127.—Lion-hunting, 128.—Disappearance of Lions, 130, 131.—Human Prey, 132.—Maneless Lions of Guzerat, 134.—A Lion Family in Bengal, 135, 136.—Prickle on the Lion's Tail, 137-139.—Nineveh Lions, 139.—Lions in the Tower of London, 140, 141.—Feats with Lions, 142.—Lion-hunting in Algeria, by Jules Gerard, 144.—The Prudhoe Lions, 144.

BIRD-LIFE.

Rate at which Birds fly, 145, 146.—Air in the Bones of Birds, 146.—Flight of the Humming-bird, 147.—Colour of Birds, 148.—Song of Birds, 149.—Beauty in Animals, 150.—Insectivorous Birds, 151.—Sea-fowl Slaughter, 152.—Hooded Crow in Zetland, 154.—Brain of Birds, 154.—Danger-signals, 155.—Addison's Love of Nature, 156, 157.

BIRDS' EGGS AND NESTS.

Colours of Eggs, 158.—Bird's-nesting, 159.—Mr. Wolley, the Ornithologist, 159, 160.—European Birds of Prey, 161.—Large Eggs, 162, 163, 164.—Baya's Nest, 164.—Oriole and Tailor-bird, 165, 166.—Australian Bower-bird, 167.—Cape Swallows, 168.—"Bird Confinement," by Dr. Livingstone.

THE EPICURE'S ORTOLAN.

Origin of the Ortolan, 172; described, 173, 174; Fattening process, 175, 176.—Prodigal Epicurism, 177, 178.

TALK ABOUT TOUCANS.

Toucan family, 179.—Gould's grand Monograph, 180.—Toucans described, 180-182; Food, 183; Habits, 184.—Gould's Toucanet, 187.

ECCENTRICITIES OF PENGUINS.

Penguins on Dassent Island, 188.—Patagonian Penguins, 189.—Falkland Islands, 189.—King Penguins, 190, 191.—Darwin's Account, 192.—Webster's Account, 193.—Swainson's Account, 194.

PELICANS AND CORMORANTS.

Pelicans described by various Naturalists, 195, 196.—The Pelican Island, 197.—Popular Error, 199-200.—Cormorants, and Fishing with Cormorants, 201-204.

TALKING BIRDS, INSTINCTS, ETC.

Sounds by various Birds, 204.—Umbrella Bird, 206.—Bittern, 207.—Butcher-bird and Parrots, 208.—Wild Swan, Laughing Goose, Cuckoo, and Nightingale, 209.—Talking Canaries, 210.—Neighing Snipe, 213.—Trochilos and Crocodiles, 216.—Instinct, Intelligence, and Reason in Birds, 217-219.—Songs of Birds and Seasons of the Day, 219.

OWLS.

Characteristics of the Owl, 221.—Owl in Poetry, 222.—Bischacho or Coquimbo, 224.—Waterton on Owls, 225, 226.—Owls, Varieties of, 227-230.

WEATHER-WISE ANIMALS.

Atmospheric Changes, 231.—Stormy Petrel, 233.—Wild Geese and Ducks, 235.—Frogs and Snails, 237.—The Mole, 240.—List of Animals, by Forster, the Meteorologist, 241.—Weatherproof Nests, 247.—"Signs of Rain," by Darwin, 248.—Shepherd of Banbury, 249.

FISH-TALK.

How Fishes Swim, 250.—Fish Changing Colour, 251.—"Fish Noise," 252.—Hearing of Fish, 253.—The Carp at Fontainebleau, 254, 255.—Affection of Fishes, 256.—Cat-fish, Anecdote of, 257.—Great Number of Fishes, 258.—Little Fishes Eaten by Medusæ, 259.—Migration of Fishes, 261.—Enormous Grampus, 262.—Bonita and Flying-fish, 263.—Jaculator Fish of Java, 264.—Port Royal, Jamaica Fish, 266.—The Shark, 267.—California, Fish of, 268.—Wonderful Fish, 269.—Vast Sunfish, 271.—Double Fish, 272.—The Square-browed Malthe, 274.—Gold Fish, 275.—The Miller's Thumb, 276.—Sea-fish Observatory, 276.—Herring Question, 278.—Aristotle's History of Animals, 279-280.

FISH IN BRITISH COLOMBIA.

Salmon-swarming, 281.—Candle-fish, 282.—Octopus, the, 283.—Sturgeon and Sturgeon Fishing, 283-287.

THE TREE-CLIMBING CRAB.

Locomotion of Fishes, 288.—Climbing Perch, 288.—Crabs in the West Indies, 289.—Crabs, Varieties of, 289-292.—Robber and Cocoa-nut Crab, 292-301.—Fish of the China Seas, 301.

MUSICAL LIZARDS.

Lizard from Formosa Isle, 303.—Its Habits, 304-306.

CHAMELEONS AND THEIR CHANGES.

The Chameleon described by Aristotle and Calmet, 307, 308.—Change of Colour, 309.—Reproduction of, 310, 311.—Tongue, 311.—Lives in Trees, 312.—Theory of Colours, 313.—The Puzzle Solved, 315.—Mrs. Belzoni's Chameleons, 317.—Lady Cust's Chameleons, 321.—Chameleon's Antipathy to Black, 322.

RUNNING TOADS.

Dr. Husenbeth's Toads at Cossey, 327.—Frog and Toad Concerts, 327.

SONG OF THE CICADA.

Greeks' Love for the Song, 329.—Cicada in British Colombia, 329.—Tennyson and Keats on the Grasshopper, 330.

STORIES ABOUT THE BARNACLE GOOSE.

Baptista Porta's Account, 331.—Max Müller on, 331.—Gerarde's Account, 332.—Giraldus Cambrensis, 332.—Professor Rolleston, Drayton's *Poly-olbion*, 333.—Sir Kenelm Digby and Sir J. Emerson Tennent, 334.—Finding the Barnacle, 334.

LEAVES ABOUT BOOKWORMS.

Bookworms, their Destructiveness, 336, 337.—How to Destroy, 338.—The Death-watch, 339.—Lines by Swift, 340.

BORING MARINE ANIMALS, AND HUMAN ENGINEERS.

Life and Labours of the Pholas, 341.—Family of the Pholas, 342.—Curious Controversy, 343.—Boring Apparatus, 342.—Several Observers, 347, 348.—Boring Annelids, 348.

LIST OF ILLUSTRATIONS.

	Page
KING PENGUINS	Frontispiece
THE TWO-HORNED AFRICAN RHINOCEROS	28
SEAL AND MERMAID	40
THE GREAT ANT-BEAR (ZOOLOGICAL SOCIETY'S)	76
FRASER'S EAGLE OWL, FROM FERNANDO PO	228
SQUARE-BROWED MALTHE AND DOUBLE FISH	274
THE TREE-CLIMBING CRAB	288
CHAMELEONS	318

ECCENTRICITIES

OF

THE ANIMAL CREATION.

INTRODUCTORY.—CURIOSITIES OF ZOOLOGY.

CURIOUS creatures of Animal Life have been objects of interest to mankind in all ages and countries; the universality of which may be traced to that feeling which "makes the whole world kin."

It has been remarked with emphatic truth by a popular writer, that "we have in the Bible and in the engraven and pictorial records the earliest evidence of the attention paid to Natural History in general. The 'navy of Tarshish' contributed to the wisdom of him who not only 'spake of the trees from the cedar of Lebanon, even unto the hyssop that springeth out of the wall,' but 'also of beasts, and of fowls, and of creeping things, and of fishes,'* to

* 1 Kings iv. 10.

say nothing of numerous other passages showing the progress that zoological knowledge had already made. The Egyptian records bear testimony to a familiarity not only with the forms of a multitude of wild animals, but with their habits and geographical distribution."

The collections of living animals, now popularly known as Zoological Gardens, are of considerable antiquity. We read of such gardens in China as far back as 2,000 years; but they consisted chiefly of some favourite animals, such as stags, fish, and tortoises. The Greeks, under Pericles, introduced peacocks in large numbers from India. The Romans had their elephants; and the first giraffe in Rome, under Cæsar, was as great an event in the history of zoological gardens at its time as the arrival in 1849 of the Hippopotamus was in London. The first zoological garden of which we have any detailed account is that in the reign of the Chinese Emperor, Wen Wang, founded by him about 1150 A.D., and named by him "The Park of Intelligence;" it contained mammalia, birds, fish, and amphibia. The zoological gardens of former times served their masters occasionally as hunting-grounds. This was constantly the case in Persia; and in Germany, so late as 1576, the Emperor Maximilian II. kept such a park for different animals near his castle, Neugebah, in which he frequently chased.

Alexander the Great possessed his zoological gardens. We find from Pliny that Alexander had given orders to the keepers to send all the rare and curious animals which died in the gardens to Aristotle.

Splendid must have been the zoological gardens which the Spaniards found connected with the Palace of Montezuma. The letters of Ferdinand Cortez and other writings of the time, as well as more recently "The History of the Indians," by Antonio Herrera, give most interesting and detailed accounts of the menagerie in Montezuma's park. The buildings belonging to these gardens were all gorgeous, as became the grandeur of the Indian prince; they were supported by pillars, each of which was hewn out of a single piece of some precious stone. Cool, arched galleries led into the different parts of the garden—to the marine and fresh-water basins, containing innumerable water-fowl,—to the birds of prey, falcons and eagles, which latter especially were represented in the greatest variety,—to the crocodiles, alligators, and serpents, some of them belonging to the most venomous species. The halls of a large square building contained the dens of the lions, tigers, leopards, bears, wolves, and other wild animals. Three hundred slaves were employed in the gardens tending the animals, upon which great care was bestowed, and scrupulous attention paid to their cleanliness. To this South American zoological garden of the sixteenth century no other of its time could be compared.*

More than six centuries ago, our Plantagenet

---

* "Athenæum."

kings kept in the Tower of London exotic animals for their recreation. The Lion Tower was built here by Henry III., who commenced assembling here a menagerie with three leopards sent to him by the Emperor Frederic II., "in token of his regal shield of arms, wherein those leopards were pictured." Here, in 1255, the Sheriffs built a house "for the King's elephant," brought from France, and the first seen in England. Our early sovereigns had a mews in the Tower as well as a menagerie :—

"Merry Margaret, as Midsomer flowre,
Gentyll as faucon and hawke of the Towre."—*Skelton.*

In the reign of Charles I., a sort of Royal Menagerie took the place of the deer with which St. James's Park was stocked in the days of Henry VIII. and Queen Elizabeth. Charles II. greatly enlarged and improved the Park; and here he might be seen playing with his dogs and feeding his ducks. The Bird-cage Walk, on the south side of the Park, had in Charles's time the cages of an aviary disposed among the trees. Near the east end of a canal was the Decoy, where water-fowl were kept; and here was Duck Island, with its salaried Governor.

Evelyn, in 1664, went to "the Physique Garden in St. James's," where he first saw "orange trees and other fine trees." He enumerates in the menagerie, "an ornocratylus, or pelican; a fowle between a storke and a swan; a melancholy water-fowl, brought from Astracan by the Russian ambassador; a milk-white raven; two Balearian cranes," one of which had a wooden leg "made by a soulder:" there were also "deere of severall countries, white, spotted like leopards; antelopes, an elk, red deer, roebucks, staggs, Guinea goates, Arabian sheepe, &c." There were "withy-potts, or nests, for the wild fowle to lay their eggs in, a little above ye surface of ye water."

"25 Feb. 1664. This night I walk'd into St. James his Parke, where I saw many strange creatures, as divers sorts of outlandish deer, Guiny sheep, a white raven, a great parrot, a storke. . . . Here are very stately walkes set with lime trees on both sides, and a fine pallmall." *

Upon the eastern island is the Swiss Cottage of the Ornithological Society, built in 1841 with a grant of 300*l.* from the Lords of the Treasury: it contains a council-room, keepers' apartments, steam-hatching apparatus; contiguous are feeding-places and decoys; and the aquatic fowl breed on the island, making their own nests among the shrubs and grasses.

The majority of Zoological Gardens now in existence have been founded in this century, with the exception of the Jardin des Plantes, which, although founded in 1626, did not receive its first living animals until the year 1793-1794. Hitherto, it had been a Garden of Plants exclusively.

We shall not be expected to enumerate the great Continental gardens, of which that at Berlin, half an hour's drive beyond the Brandenburg gates, contains the Royal Menagerie; it is open upon the payment of an admission fee, and generally resembles

* Journal of Mr. E. Browne, son of Sir Thomas Browne.

our garden at the Regent's Park. Berlin has also its Zoological Collection in its Museum of Natural History. This collection is one of the richest and most extensive in Europe, especially in the department of Ornithology: it includes the birds collected by Pallas and Wildenow, and the fishes of Bloch. The best specimens are those from Mexico, the Red Sea, and the Cape. The whole is exceedingly well arranged, and *named* for the convenience of students. Still, our Zoological Collection in the British Museum (to be hereafter removed to South Kensington) is allowed to be the finest in Europe.

The Zoological Society of London was instituted in 1826, and occupies now about seventeen acres of gardens in the Regent's Park. Among the earliest tenants of the Menagerie were a pair of emues from New Holland; two Arctic bears and a Russian bear; a herd of kangaroos; Cuban mastiffs and Thibet watch-dogs; two llamas from Peru; a splendid collection of eagles, falcons, and owls; a pair of beavers; cranes, spoonbills, and storks; zebras and Indian cows; Esquimaux dogs; armadilloes; and a collection of monkeys. To the menagerie have since been added an immense number of species of *Mammalia* and *Birds;* in 1849, a collection of *Reptiles;* and in 1853, a collection of *Fish, Mollusca, Zoophytes,* and other *Aquatic Animals.* In 1830, the menagerie collected by George IV. at Sandpit-gate, Windsor, was removed to the Society's Gardens; and 1834 the last of the Tower Menagerie was received here. It is now the finest public Vivarium in Europe.

The following are some of the more remarkable

animals which the Society have possessed, or are now in the menagerie:—

Antelopes, the great family of, finely represented. The beautiful *Elands* were bequeathed by the late Earl of Derby, and have bred freely since their arrival in 1851. The Leucoryx is the first of her race born out of Africa. *Anteater, Giant*, brought to England from Brazil in 1853, was exhibited in Broad-street, St. Giles's, until purchased by the Zoological Society for 200*l*. *Apteryx*, or *Kiwi* bird, from New Zealand; the first living specimen brought to England of this rare bird. The *Fish-house*, built of iron and glass, in 1853, consisting of a series of glass tanks, in which fish spawn, zoophytes produce young, and algæ luxuriate; crustacea and mollusca live successfully, and ascidian polypes are illustrated, together with sea anemones, jelly-fishes, and star-fishes, rare shell-fishes, &c.: a new world of animal life is here seen as in the depths of the ocean, with masses of rock, sand, gravel, corallines, sea-weed, and sea-water; the animals are in a state of natural restlessness, now quiescent, now eating and being eaten. *Aurochs*, or *European Bisons:* a pair presented by the Emperor of Russia, in 1847, from the forest of Bialowitzca: the male died in 1848, the female in 1849, from pleuro-pneumonia. *Bears:* the collection is one of the largest ever made. *Elephants:* including an Indian elephant calf and its mother. In 1847 died here the great Indian elephant Jack, having been in the gardens sixteen years. Adjoining the stable is a tank of water, of a depth nearly equal to the height of a full-grown elephant. In 1851 the Society possessed a *herd of four elephants*, besides a hippopotamus, a rhinoceros, and both species of tapir; being the largest collection of pachydermata ever exhibited in Europe. *Giraffes:* four received in 1836 cost the Society upwards of 2,300*l*., including 1,000*l*. for steamboat passage: the female produced six male fawns here between 1840 and 1851. *Hippopotamus*, a young male (the first living specimen seen in England), received from Egypt in May, 1850, when ten months old, seven feet long, and six and a-half feet in girth; also a female hippopotamus, received 1854. *Humming-birds:* Mr. Gould's matchless collection of 2,000 examples was exhibited here in 1851 and 1852. *Iguanas,* two from Cuba and Carthagena, closely resembling, in everything but size, the fossil Iguanodon. The *Lions* number generally from eight to ten, including a pair of cubs born in the gardens in 1853. *Orang-utan* and *Chimpanzee:* the purchase-money of the latter sometimes exceeds 300*l.* The orang "Darby," brought from Borneo in 1851, is the finest yet seen in Europe, very intelligent, and docile as a child. *Parrot-houses,* the, sometimes contain from sixty to seventy species. *Rapacious Birds:* so extensive a series of eagles and vultures has never yet been seen at one view. The *Reptile-house* was fitted up in 1849; the creatures are placed in large plate-glass cases: here are pythons and a rattle-snake, with a young one born here; here is also a case of the tree-frogs of Europe: a yellow snake from Jamaica has produced eight young in the gardens. *Cobra de Capello,* from India: in 1852, a keeper in the gardens was killed by the bite of this serpent. *A large Boa* in 1850 swallowed a blanket, and disgorged it in thirty-three days. A *one-horned Rhinoceros,* of continental India, was obtained in 1834, when it was about four years old, and weighed 26 cwt.; it died in 1850: it was replaced by a female, about five years old. *Satin Bower-Birds,* from Sydney: a pair have built here a bower, or breeding-place. *Tapir* of the Old World, from Mount Ophir; the nearest existing form the Paleotherium. *Tigers:* a pair of magnificent specimens, presented by the Guicoway of Baroda in 1851; a pair of clouded tigers, 1854. The *Wapiti Deer* breeds every year in the Menagerie.

The animals in the Gardens, although reduced in number, are more valuable and interesting than when their number was higher. The mission of the Society's head-keeper, to collect rare animals for the Menagerie, has been very profitable. The additional houses from time to time, are very expensive: the new monkey house, fittings, and work cost 4,842*l.*; and in 1864, the sum of 6,604*l.* was laid out in permanent additions to the establishment.

Very rare, and consequently expensive, animals are generally purchased. Thus, the first Rhinoceros cost 1,000*l.*; the four Giraffes, 700*l.*, and their carriage an additional 700*l.* The Elephant and calf were bought in 1851 for 500*l.*; and the Hippopotamus, although a gift, was not brought home and housed at less than 1,000*l.*—a sum which he more than realised in the famous Exhibition season, when the receipts were 10,000*l.* above the previous year. The Lion Albert was purchased for 140*l.*; a tiger, in 1852, for 200*l.* The value of some of the smaller birds will appear, however, more startling : thus, the pair of black-necked Swans were purchased for 80*l.*; a pair of crowned Pigeons and two Maleos, 60*l.*; a pair of Victoria Pigeons, 35*l.*; four Mandarin Ducks, 70*l.* Most of these rare birds (now in the great aviary) came from the Knowsley collection, at the sale of which, in 1851, purchases were made to the extent of 985*l.* It would be impossible from these prices, however, to judge of the present value of the animals. Take the Rhinoceros, for example : the first specimen cost 1,000*l.*; the second, quite as fine a brute, only 350*l.* Lions range again from 40*l.* to 180*l.*, and Tigers from 40*l.* to 200*l.* The ignorance displayed by some persons as to the value of well-known objects is something marvellous.

—A sea-captain demanded 600*l*. for a pair of Pythons, and at last took 40*l*.! An American offered the Society a Grisly Bear for 2,000*l*., to be delivered in the United States; and, more laughable still, a moribund Walrus, which had been fed for nine weeks on salt pork and meal, was offered for the trifling sum of 700*l*.!

There is a strange notion that the Zoological Society has proposed a large reward for a "Tortoiseshell Tom-cat," and one was accordingly offered to the Society for 250*l*.! But male Tortoiseshell Cats may be had in many quarters.*

The Surrey Zoological Gardens were established in 1831. Thither Cross removed his menagerie from the King's Mews, where it had been transferred from

* In April, 1842, Mr. Batty's collection of animals was sold by auction, when the undermentioned animals brought— Large red-faced Monkey (clever), 1*l*. 10*s*.; fine Coatimondi, 1*l*. 4*s*.; Mandril (the only one in England), 1*l*. 17*s*.; pair of Java Hares, 1*l*. 9*s*.; a Puma, 14*l*.; handsome Senegal Lioness, 9*l*.; a Hyæna, 7*l*.; splendid Barbary Lioness, 24*l*.; handsome Bengal Tigress, 90*l*.; brown Bear, 6*l*.; the largest Polar Bear in Europe, 37*l*.; pair of Esquimaux Sledge-Dogs, 3*l*. 7*s*.; pair of Golden Pheasants, 3*l*. 10*s*.; a blue-and-buff Macaw (clever talker), 2*l*. 10*s*.; a horned Owl, from North America, 3*l*. 10*s*.; a magnificent Barbary Lion, trained for performance, 105 guineas; a Lioness, similarly trained, 90 guineas; handsome Senegal performing Leopard, 34 guineas; two others, 50 guineas; Ursine Sloth, 12 guineas; Indian Buffalo, 10 guineas; sagacious male Elephant, trained for theatrical performances, 350 guineas. The above is stated to have been the first sale of the kind by public auction in this country.

Exeter Change. At Walworth a glazed circular building, 100 feet in diameter, was built for the cages of the carnivorous animals (Lions, Tigers, Leopards, &c.); and other houses for Mammalia, Birds, &c. Here, in 1834, was first exhibited a young Indian one-horned Rhinoceros, for which Cross paid 800*l*. It was the only specimen brought to England for twenty years. In 1836 were added three Giraffes, one fifteen feet high. The menagerie was dispersed in 1856. The menagerie at Exeter Change was a poor collection, though the admission-charge was, at one period, half-a-crown!

The collections of animals exhibited at fairs have added little to Zoological information; but we may mention that Wombwell, one of the most noted of the showfolk, bought a pair of the first Boa Constrictors imported into England: for these he paid 75*l*., and in three weeks realized considerably more than that sum by their exhibition. At the time of his death, in 1850, Wombwell was possessed of three huge menageries, the cost of maintaining which averaged at least 35*l*. per day; and he used to estimate that, from mortality and disease, he had lost, from first to last, from 12,000*l*. to 15,000*l*.

Our object in the following succession of sketches of the habits and eccentricities of the more striking animals, and their principal claims upon our attention, is to present, in narrative, their leading characteristics, and thus to secure a willing audience from old and young.

THE RHINOCEROS IN ENGLAND.

THE intellectual helps to the study of zoology are nowhere more strikingly evident than in the finest collection of pachyderms (thick-skinned animals) in the world, now possessed by our Zoological Society. Here we have a pair of Indian Elephants, a pair of African Elephants, a pair of Hippopotami, a pair of Indian Rhinoceroses, and an African or two-horned Rhinoceros.

The specimens of the Rhinoceros which have been exhibited in Europe since the revival of literature have been few and far between. The first was of the one-horned species, sent from India to Emmanuel, King of Portugal, in the year 1513. The Sovereign made a present of it to the Pope; but the animal being seized during its passage with a fit of fury, occasioned the loss of the vessel in which it was transported. A second Rhinoceros was brought to England in 1685; a third was exhibited over almost the whole of Europe in 1739; and a fourth, a female, in 1741. A fifth specimen arrived at Versailles in

1771, and it died in 1793, at the age of about twenty-six years. The sixth was a very young Rhinoceros, which died in this country in the year 1800. The seventh, a young specimen, was in the possession of Mr. Cross, at Exeter Change, about 1814 ; and an eighth specimen was living about the same time in the Garden of Plants at Paris. In 1834 Mr. Cross received at the Surrey Gardens, from the Birman empire, a Rhinoceros, a year and a-half old, as already stated at page 21. In 1851 the Zoological Society purchased a full-grown female Rhinoceros; and in 1864 they received a male Rhinoceros from Calcutta. All these specimens were from India, and *one-horned;* so that the *two-horned* Rhinoceros had not been brought to England until the arrival of an African Rhinoceros, *two-horned*, in September, 1868.*

The ancient history of the Rhinoceros is interesting, but intricate. It seems to be mentioned in several passages of the Scriptures, in most of which the animal or animals intended to be designated was or were the *Rhinoceros unicornis,* or Great Asiatic one-horned Rhinoceros. M. Lesson expresses a decided opinion to this effect: indeed, the description in Job (chap. xxxix.) would almost forbid the

---

* The conveyance of a Rhinoceros over sea is a labour of some risk. In 1814 a full-grown specimen on his voyage from Calcutta to this country became so furious that he was fastened down to the ship's deck, with part of a chain-cable round his neck; and even then he succeeded in destroying a portion of the vessel, till, a heavy storm coming on, the Rhinoceros was thrown overboard to prevent the serious consequence of his getting loose in the ship.

conclusion that any animal was in the writer's mind except one of surpassing bulk and indomitable strength. The impotence of man is finely contrasted with the might of the Rhinoceros in this description, which would be overcharged if it applied to the less powerful animals alluded to in the previous passages.

It has also been doubted whether accounts of the Indian Wild Ass, given by Ctesias, were not highly coloured and exaggerated descriptions of this genus; and whether the Indian Ass of Aristotle was not a Rhinoceros.

Agatharchides describes the one-horned Rhinoceros by name, and speaks of its ripping up the belly of the Elephant. This is, probably, the earliest occurrence of the name *Rhinoceros*. The Rhinoceros which figured in the celebrated pomps of Ptolemy Philadelphus was an Ethiopian, and seems to have marched last in the procession of wild animals, probably on account of its superior rarity, and immediately after the Cameleopard.

Dion Cassius speaks of the Rhinoceros killed in the circus with a Hippopotamus in the show given by Augustus to celebrate his victory over Cleopatra; he says that the Hippopotamus and this animal were then first seen and killed at Rome. The Rhinoceros then slain is thought to have been African, and two-horned.

The Rhinoceros clearly described by Strabo, as seen by him, was one-horned. That noticed by Pausanias as "the Bull of Ethiopia," was two-horned, and he describes the relative position of the horns.

Wood, in his "Zoography," gives an engraving of the coin of Domitian (small Roman brass), on the reverse of which is the distinct form of a two-horned Rhinoceros: its exhibition to the Roman people, probably of the very animal represented on the coin, is particularly described in one of the epigrams attributed to Martial, who lived in the reigns of Titus and Domitian. By the description of the epigram it appears that a combat between a Rhinoceros and a Bear was intended, but that it was very difficult to irritate the more unwieldy animal so as to make him display his usual ferocity; at length, however, he tossed the bear from his double horn, with as much facility as a bull tosses to the sky the bundles placed for the purpose of enraging him. Thus far the coin and the epigram perfectly agree as to the existence of the double horn; but, unfortunately, commentators and antiquaries were not to be convinced that a Rhinoceros could have more than one horn, and have at once displayed their sagacity and incredulity in their explanations on the subject.

Two, at least, of the two-horned Rhinoceroses were shown at Rome in the reign of Domitian. The Emperors Antoninus, Heliogabalus, and Gordian also exhibited Rhinoceroses. Cosmas speaks expressly of the Ethiopian Rhinoceros as having two horns, and of its power of moving them.

The tractability of the Asiatic Rhinoceros has been confirmed by observers in the native country of the animal. Bishop Heber saw at Lucknow five or six very large Rhinoceroses, of which he found that prints and drawings had given him a very imperfect conception. They were more bulky animals, and of a darker colour than the Bishop supposed; though the latter difference might be occasioned by oiling the skin. The folds of their skin also surpassed all which the Bishop had expected. Those at Lucknow were quiet and gentle animals, except that one of them had a feud with horses. They had sometimes howdahs, or chaise-like seats, on their backs, and were once fastened in a carriage, but only as an experiment, which was not followed up. The Bishop, however, subsequently saw a Rhinoceros (the present of Lord Amherst to the Guicwar), which was so tamed as to be ridden by a Mohout quite as patiently as an elephant.

No two-horned Rhinoceros seems to have been brought alive to Europe in modern times. Indeed, up to a comparatively late period, their form was known only by the horns which were preserved in museums; nor did voyagers give any sufficient details to impart any clear idea of the form of the animal. The rude figure given by Aldrovandus, in 1639, leaves no doubt that, wretched as it is, it must have been taken from a two-horned Rhinoceros.

Dr. Parsons endeavoured to show that the one-horned Rhinoceros always belonged to Asia, and the two-horned Rhinoceros to Africa; but there are two-horned Rhinoceroses in Asia, as well as in Africa. Flacourt saw one in the Bay of Soldaque, near the Cape of Good Hope, at a distance. Kolbe and others always considered the Rhinoceros of the Cape as two-horned; but Colonel Gordon seems to be the first who entirely detailed the species with any exactness. Sparrman described the Cape Rhinoceros, though his figure of the animal is stiff and ill-drawn. At this period it was well known that the Cape species was not only distinguished by having two horns from the Indian Rhinoceros then known, but also by an absence of the folds of the skin so remarkable in the latter.

We should here notice the carelessness, to call it by the mildest name, of Bruce, who gave to the world a representation of a two-horned Rhinoceros from Abyssinia, with a strongly folded skin. The truth appears to be that the body of the animal figured by Bruce was copied from that of the one-horned Rhinoceros given by Buffon, to which Bruce added a second horn. Salt proved that the Abyssinian Rhinoceros is two-horned, and that it resembles that of the Cape.

Sparmann exposes the errors and poetic fancies of Buffon respecting the impenetrable nature of the skin. He ordered one of his Hottentots to make a trial of this with his hassagai on a Rhinoceros which had been shot. Though this weapon was far from being in good order, and had no other sharpness than that which it had received from the forge, the Hottentot, at the distance of five or six paces, not only pierced with it the thick hide of the animal, but buried it half a foot deep in its body.

Mr. Tegetmeier has sufficiently described in the "Field" journal the African Rhinoceros just received at the Zoological Society's menagerie in the Regent's-park, and which has been sketched by Mr. T. W. Wood expressly for the present volume.

It was captured about a year ago in Upper Nubia by the native hunters in the employment of Mr. Casanova, at Kassala; and was sent, by way of Alexandria and Trieste, to Mr. Karl Hagenbeck, of Hamburg, a dealer in wild beasts, who sold it to the Zoological Society.

"This animal is very distinct from its Asiatic congeners; it differs strikingly in the number of horns, as well as in the character of its skin, which is destitute of those large folds, which cause the Indian species to remind the observer of a gigantic 'hog in armour.'

"The arrival of this animal will tend to clear up the confusion that prevails respecting the number of distinct species of African Rhinoceros. Some writers—as Sir W. C. Harris—admit the existence of two species only, the dark and the light, or, as they are termed, the 'white' and the 'black.' Others, as Dr. A. Smith, describe three; some, as the late Mr. Anderssen, write of four; and Mr. Chapman even speaks of a fifth species or hybrid.

"Three of these species are very distinctly defined—the ordinary dark animal, the *Rhinoceros bicornis*, in which the posterior horn is much shorter than the anterior; the *Rhinoceros keitloa*, in which the two horns are of equal length; and the 'white' species, *Rhinoceros simus*. The last, among other characters, is, according to Dr. Smith,

distinguished by the square character of the upper lip, which is not prehensile.

"The young animal now (October, 1868) in the Zoological Society's garden, appears to belong to the first-named species, the largest specimens of which when full grown reach a height of 6ft., and a length of 13ft., the tail not included. Its present height is 3½ft., and length about 6ft. In general appearance the mature animal resembles a gigantic pig, the limbs being brought under the body. The feet are most singular in form, being very distinctly three-toed, and the remarkable trefoil-like *spoors* that they make in the soil render the animal easy to track. The horns vary greatly in length in different animals; the first not unfrequently reaches a length of 2ft., the second being considerably shorter. These appendages differ very much from ordinary horns; they are, in fact, more of the nature of agglutinated hair, being attached to the skin only, and consequently they separate from the skull when the latter is preserved.

"The head is not remarkable for comeliness, especially in the mature animal, in which the skin of the face is deeply wrinkled, and the small eyes are surrounded with many folds. The upper lip is elongated, and is used in gathering the food. The adult animals are described by Sir W. C. Harris, in his 'Illustrations of the Game Animals of South Africa,' as 'swinish, cross-grained, ill-tempered, wallowing brutes.'"

Mr. Burchell, during his travels in Africa, shot

nine Rhinoceroses, besides a smaller one. The latter he presented to the British Museum. The animal is, however, becoming every day more and more scarce in Southern Africa; indeed, it is rarely to be met with in some parts. It appears that, in one day, two Rhinoceroses were shot by Speelman, the faithful Hottentot who attended Mr. Burchell. He fired off his gun but twice, and each time he killed a Rhinoceros! The animal's sense of hearing is very quick: should he be disturbed, he sometimes becomes furious, and pursues his enemy; and then, if once he gets sight of the hunter, it is scarcely possible for him to escape, unless he possesses extraordinary coolness and presence of mind. Yet, if he will quietly wait till the enraged animal makes a run at him, and will then spring suddenly on one side, to let it pass, he may gain time enough for reloading his gun before the Rhinoceros gets sight of him again, which, fortunately, owing to its imperfection of sight, it does slowly and with difficulty.

Speelman, in shooting a large male Rhinoceros, used bullets cast with an admixture of tin, to render them harder. They were flattened and beat out of shape by striking against the bones, but those which were found lodged in the fleshy parts had preserved their proper form, a fact which shows how little the hardness of the creature's hide corresponds with the vulgar opinion of its being impenetrable to a musketball. Mr. Burchell found this Rhinoceros nearly cut up. On each side of the carcase the Hottentots had made a fire to warm themselves; and round a third fire were assembled at least twenty-four Bushmen, most of whom were employed the whole night long in broiling, eating, and talking. Their appetite seemed insatiable, for no sooner had they broiled and eaten one slice of meat than they turned to the carcase and cut another. The meat was excellent, and had much the taste of beef. "The tongue," says Mr. Burchell, "is a dainty treat, even for an epicure." The hide is cut into strips, three feet or more in length, rounded to the thickness of a man's finger, and tapering to the top. This is called a *shambok*, and is universally used in the colony of the Cape for a horsewhip, and is much more durable than the whips of European manufacture. The natural food of the Rhinoceros, till the animal fled before the colonists, was a pale, bushy shrub, called the Rhinoceros-bush, which burns while green as freely as the driest fuel, so as readily to make a roadside fire.

The horn of the Rhinoceros, single or double, has its special history by the way of popular tradition. From the earliest times this horn has been supposed to possess preservative virtues and mysterious properties—to be capable of curing diseases and discovering the presence of poison; and in all countries where the Rhinoceros exists, but especially in the East, such is still the opinion respecting it. In the details of the first voyage of the English to India, in 1591, we find Rhinoceros' horns monopolised by the native sovereigns on account of their reputed virtues in detecting the presence of poison.

Thunberg observes, in his "Journey into Caffraria," that "the horns of the Rhinoceros were kept by some people, both in town and country, not only as rarities, but also as useful in diseases, and for the purpose of detecting poisons. As to the former of these intentions, the fine shavings were supposed to cure convulsions and spasms in children. With respect to the latter, it was generally believed that goblets made of these horns would discover a poisonous draught that was poured into them, by making the liquor ferment till it ran quite out of the goblet. Of these horns goblets are made, which are set in gold and silver and presented to kings, persons of distinction, and particular friends, or else sold at a high price, sometimes at the rate of fifty rix-dollars each." Thunberg adds:—"When I tried these horns, both wrought and unwrought, both old and young horns, with several sorts of poison, weak as well as strong, I observed not the least motion or effervescence; but when a solution of corrosive sublimate or other similar substance was poured into one of these horns, there arose only a few bubbles, produced by the air which had been enclosed in the pores of the horn and which were now disengaged."

Rankin (in his "Wars and Sports") states this mode of using it: a small quantity of water is put into the concave part of the root, then hold it with the point downwards and stir the water with the point of an iron nail till it is discoloured, when the patient is to drink it.

STORIES OF MERMAIDS.

LESS than half a century ago, a pretended Mermaid was one of the sights of a London season; to see which credulous persons rushed to pay half-crowns and shillings with a readiness which seemed to rebuke the record—that the existence of a Mermaid is an exploded fallacy of two centuries since.

Mermaids have had a legendary existence from very early ages, for the Sirens of the ancients evidently belonged to the same remarkable family. Shakspeare uses the term Mermaid as synonymous with Siren:—

> "O train me not, sweet Mermaid, with thy note,
> To drown me in thy sister's flood of tears;
> Sing, Syren, for thyself."—*Comedy of Errors*, iii. 2.

Elsewhere, Shakspeare's use of the term is more applicable to the Siren than to the common idea of a Mermaid; as in the "Midsummer'Night's Dream," where the "Mermaid on a dolphin's back" could not easily have been so placed. A Merman, the male of this imaginary species, is mentioned by Taylor, the water-poet:—

> "A thing turmoyling in the sea we spide,
> Like to a Meareman."

An old writer has this ingenious illustration:— "Mermaids, in Homer, were witches, and their songs enchantments;" which reminds us of the invitation in Haydn's Mermaid's Song:—

> "Come with me, and we will go
> Where the rocks of coral grow."

The orthodox Mermaid is half woman, half fish; and the fishy half is sometimes depicted as doubly tailed, such as we see in the heraldry of France and Germany; and in the Basle edition of Ptolemy's "Geography," dated 1540, a double-tailed Mermaid figures in one of the plates. In the arms of the Fishmongers' Company of London, the supporters are "a Merman and maid, first, armed, the latter with a mirror in the left hand, proper." From this heraldic employment, the Mermaid became a popular tavern sign; and there was an old dance called the Mermaid.

Sir Thomas Browne refers to the *picture* of Mermaids, though he does not admit their existence. They "are conceived to answer the shape of the ancient Sirens that attempted upon Ulysses; which, notwithstanding, were of another description, containing no fishy composure, but made up of man and bird." Sir Thomas is inclined to refer the Mermaid to Dagon, the tutelary deity of the Philistines, which, according to the common opinion, had a human female bust and a fish-like termination; though the details of this fish idolatry are entirely conjectural.

Leyden, the Scottish poet, has left a charming ballad, entitled "The Mermaid," the scene of which is laid at Corrievreckin: the opening of this poem Sir Walter Scott praised as exhibiting a power of numbers which, for mere melody of sound, has seldom been excelled in English poetry:—

> "On Jura's heath how sweetly swell
>   The murmurs of the mountain bee!
> How softly mourns the writhèd shell
>   Of Jura's shore its parent sea!
>
> "But softer floating, o'er the deep,
>   The Mermaid's sweet sea-soothing lay,
> That charmed the dancing waves to sleep
>   Before the bark of Colonsay."

The ballad thus describes the wooing of the gallant chieftain:—

> "Proud swells her heart! she deems at last
>   To lure him with her silver tongue,
> And, as the shelving rocks she passed,
>   She raised her voice, and sweetly sung.
>
> "In softer, sweeter strains she sung,
>   Slow gliding o'er the moonlight bay,
> When light to land the chieftain sprung,
>   To hail the maid of Colonsay.
>
> "O sad the Mermaid's gay notes fell,
>   And sadly sink remote at sea!
> O sadly mourns the writhèd shell
>   Of Jura's shore, its parent sea.

> "And ever as the year returns,
> The charm-bound sailors know the day;
> For sadly still the Mermaid mourns
> The lovely chief of Colonsay."

Curious evidences of the existence of Mermaids are to be found in ancient authors. Pliny says that "the ambassadors to Augustine from Gaul declared that sea-women were often seen in their neighbourhood." Solinus and Aulus Gellius also speak of their existence. Some stories are, however, past credence. It is related in the "Histoire d'Angleterre" that, in the year 1187, a Merman was "fished up" off the coast of Suffolk, and kept for six months. It was like a man, but wanted speech, and at length escaped into the sea! In 1430, in the great tempests which destroyed the dykes in Holland, some women at Edam, in West Friesland, saw a Mermaid who had been driven by the waters into the meadows, which were overflowed. "They took it, dressed it in female attire, and taught it to spin!" It was taken to Haarlem, where it lived some years! Then we read of Ceylonese fishermen, in 1560, catching, at one draught, seven Mermen and Mermaids, which were dissected! In 1531, a Mermaid, caught in the Baltic, was sent to Sigismund, King of Poland, with whom she lived three days, and was seen by the whole court!

In Merollo's "Voyage to Congo," in 1682, Mermaids are said to be plentiful all along the river Zaire. In the "Aberdeen Almanack" for 1688, it is predicted that "near the place where the famous

Dee payeth his tribute to the German Ocean," on the 1st, 13th, and 29th of May, and other specified times, curious observers may "undoubtedly see a pretty company of Mar Maids," and likewise hear their melodious voices. In another part of Scotland, about the same time, Brand, in his "Description of Orkney and Shetland," tells us that two fishermen drew up with a hook a Mermaid, "having face, arm, breast, shoulders, &c., of a woman, and long hair hanging down the neck, but the nether part, from below the waist, hidden in the water." One of the fishermen stabbed her with a knife, and she was seen no more! The evidence went thus:—Brand was told by a lady and gentleman, who were told by a baillie to whom the fishing-boat belonged, who was told by the fishers! Valentyn describes a Mermaid he saw in 1714, on his voyage from Batavia to Europe, "sitting on the surface of the water," &c. In 1758, a Mermaid is said to have been exhibited at the fair of St. Germain, in France. It was about two feet long, and sported about in a vessel of water. It was fed with bread and fish. It was a female, with negro features.

In 1775 appeared a very circumstantial account of a Mermaid which was captured in the Grecian Archipelago in the preceding year, and exhibited in London. The account is ludicrously minute, and it ends with : "It is said to have an enchanting voice, which it never exerts except before a storm." This imposture was craftily made up out of the skin of the angle shark. In Mr. Morgan's "Tour to Milford Haven in the year 1795," appears an equally circumstantial account of a Mermaid, said to have been seen by one Henry Reynolds, a farmer, of Ren-y-hold, in the parish of Castlemartin, in 1782. It resembled a youth of sixteen or eighteen years of age, with a very white skin : it was bathing. The evidence is very roundabout, so that there were abundant means for converting some peculiar kind of fish into a Merman, without imputing intentional dishonesty to any one. "Something akin to this kind of evidence is observable in the account of a Mermaid seen in Caithness in 1809, which attracted much attention in England as well as in Scotland, and induced the Philosophical Society of Glasgow to investigate the matter. The Editor of a newspaper, who inserted the statement, had been told by a gentleman, who had been shown a letter by Sir John Sinclair, who had obtained it from Mr. Innes, to whom it had been written by Miss Mackay, who had heard the story from the persons (two servant girls and a boy) who had seen the strange animal in the water." (Chambers's "Book of Days.")

Then we read of a so-called Mermaid, shown in the year 1794 at No. 7, Broad-court, Bow-street, Covent-garden, said to have been taken in the North Seas by Captain Foster. It was of the usual description.

Much evidence comes from Scotland. Thus, in the year 1797, a schoolmaster of Thurso affirmed that he had seen a Mermaid, apparently in the act of combing her hair with her fingers ! Twelve years afterwards, several persons observed near the same place a like appearance. Dr. Chisholm, in his "Essay on Malignant Fever in the West Indies," in 1801, relates that, in the year 1797, happening to be at Governor Van Battenburg's plantation, in Berbice, " the conversation turned on a singular animal which had been repeatedly seen in Berbice river, and some smaller rivers. This animal is the famous Mermaid, hitherto considered as a mere creature of the imagination. It is called by the Indians *méné*, mamma, or mother of the waters. The description given of it by the Governor is as follows:—'The upper portion resembles the human figure, the head smaller in proportion, sometimes bare, but oftener covered with a copious quantity of long black hair. The shoulders are broad, and the breasts large and well-formed. The lower portion resembles the tail of a fish, is of great dimensions, the tail forked, and not unlike that of the dolphin, as it is usually represented. The colour of the skin is either black or tawny.' The animal is held in veneration by the Indians, who imagine that killing it would be attended with calamitous consequences. It is from this circumstance that none of these animals have been shot, and consequently examined but at a distance. They have been generally observed in a sitting posture in the water, none of the lower extremity being seen until they are disturbed, when, by plunging, the tail agitates the water to a considerable distance round. They have been always seen employed in smoothing their hair, and have thus been frequently taken for Indian women bathing." In 1811, a young man, named John M'Isaac, of Corphine, in Kintyre, in Scotland, made oath, on examination at Campbell-town, that he saw, on the 13th of October in the above year, on a rock on the sea-coast, an animal which generally corresponded with the form of the Mermaid—the upper half human shape, the other brindled or reddish grey, apparently covered with scales; the extremity of the tail greenish red; head covered with long hair, at times put back on both sides of the head. This statement was attested by the minister of Campbell-town and the Chamberlain of Mull.

In August, 1812, Mr. Toupin, of Exmouth, in a sailing excursion, and when about a mile south-east of Exmouth Bar, heard a sound like that of the Æolian harp; and saw, at about one hundred yards distance, a creature, which was regarded as a Mermaid. The head, from the crown to the chin, formed a long oval, and the face seemed to resemble that of the seal, though with more agreeable features. The presumed hair, the arms, and the hand, with four fingers connected by a membrane, are then described, and the tail with polished scales. The entire height of the animal was from five feet to five and a-half feet. In 1819, a creature approached the coast of Ireland. It was about the size of a child ten years of age, with prominent bosom, long dark hair, and dark eyes. It was shot at, when it plunged into the sea with a loud scream.

SEAL AND MERMAID.

In reviewing these stories of Mermaids, it may be remarked that there is always a fish in each tale—either a living fish of a peculiar kind, which a fanciful person thinks to bear some resemblance in the upper part to a human being, or a fish which becomes marvellous in the progress of its description from mouth to mouth. It is commonly thought the seals may often have been mistaken for Mermaids. But, of all the animals of the whale tribe that which approaches the nearest in form to man is, undoubtedly, the Dugong, which, when its head and breast are raised above the water, and its pectoral fins, resembling hands, are visible, might easily be taken by superstitious seamen for a semi-human being, or a Mermaid. Of this deception a remarkable instance occurred in 1826. The skeleton of a Mermaid, as it was called, was brought to Portsmouth, which had been shot in the vicinity of the Island of Mombass. This was submitted to the members of the Philosophical Society, when it proved to be the skeleton of a Dugong. To those who came to the examination with preconceived notions of a fabulous Mermaid, it presented, as it lay on the lecture-table, a singular appearance. It was about six feet long; the lower portion, with its broad tail-like extremity, suggested the idea of a powerful fish-like termination, whilst the forelegs presented to the unskilful eye a resemblance to the bones of a small female arm; the cranium, however, had a brutal form, which could never have borne the lineaments of "the human face divine."

The Mermaid has been traced to the Manatee as well as to the Dugong: the former is an aquatic animal, externally resembling a whale, and named from its flipper, resembling the human hand, *manus*. Again, the *mammæ* (teats) of the Manatees and Dugongs are pectoral; and this conformation, joined to the adroit use of their flippers (whose five fingers can easily be distinguished through the inverting membranes, four of them being terminated by nails) in progression, nursing their young, &c., have caused them, when seen at a distance with the anterior part of their body out of the water, to be taken for some creature approaching to human shape so nearly (especially as their middle is thick set with hair, giving somewhat of the effect of human hair or a beard), that there can be little doubt that not a few of the tales of Mermen and Mermaids have had their origin with these animals as well as with seals and walruses. Thus the Portuguese and Spaniards give the *Manatee* a denomination which signifies Woman-fish; and the Dutch call the Dugong *Baardanetjee*, or Little-bearded Man. A very little imagination and a memory for only the marvellous portion of the appearance sufficed, doubtless, to complete the metamorphosis of this half woman or man, half fish, into a Siren, a Mermaid, or a Merman; and the wild recital of the voyager was treasured up by writers who, as Cuvier well observes, have displayed more learning than judgment.

The comb and the toilet-glass have already been incidentally mentioned as accessories in these Mermaid stories; and these, with the origin of the creature, Sir George Head thus ingeniously attempts to explain:—" The resemblance of the seal, or sea-calf, to the calf consists only in the voice, and the voice of the calf is certainly not dissimilar to that of a man. But the claws of the seal, as well as the hand, are like a lady's back-hair comb; wherefore, altogether, supposing the resplendence of sea-water streaming down its polished neck, on a sunshiny day, the substitute for a looking-glass, we arrive at once at the fabulous history of the marine maiden or mermaid, and the appendages of her toilet."

The progress of zoological science has long since destroyed the belief in the existence of the Mermaid. If its upper structure be human, with lungs resembling our own, how could such a creature live and breathe at the bottom of the sea, where it is stated to be? for our most expert divers are unable to stay under water more than half an hour. Suppose it to be of the cetaceous class, it could only remain under the water two or three minutes together without rising to the surface to take breath; and if this were the case with the Mermaid, would it not be oftener seen?

Half a century has scarcely elapsed since a *manufactured* Mermaid was shown in London with all the confidence of its being a natural creature. In the winter of 1822 there was exhibited at the Egyptian Hall, in Piccadilly, this pretended Mermaid, which was visited by from 300 to 400 persons daily! The imposture, however, was too gross to last long; and it was ascertained to be the dried skin of the head and shoulders of a monkey attached very neatly to the dried skin of a fish of the salmon kind with the head cut off; the compound figure being stuffed and highly varnished, the better to deceive the eye. This grotesque object was taken by a Dutch vessel from on board a native Malacca boat; and from the reverence shown to it by the sailors it is supposed to have represented the incarnation of one of the idol gods of the Malacca Islands. A correspondent of the "Magazine of Natural History," 1829, however, avers that the above "Mermaid" was brought from the East Indies; for being at St. Helena in 1813 he saw it on board the ship which was bringing it to England. The impression on his mind was that it was an artificial compound of the upper part of a small ape with the lower half of a fish; and by aid of a powerful glass he ascertained the point of union between the two parts. He was somewhat staggered to find that this was so neatly effected that the precise line of junction was not satisfactorily apparent: the creature was then in its best state of preservation.

In a volume of "Manners and Customs of the Japanese," published in 1841, we, however, find the following version of the history of the above Mermaid:—" A Japanese fisherman seems to have displayed ingenuity for the mere purpose of making money by his countrymen's passion for everything odd and strange. He contrived to unite the upper half of a monkey to the lower half of a fish so neatly as to defy ordinary inspection. He then gave out that he had caught the creature in his net, but that it had died shortly after being taken out of the water; and he derived considerable pecuniary profit from his cunning in more ways than one. The exhibition of the sea monster to Japanese curiosity paid well; yet more productive was the assertion that the half-human fish, having spoken during the five minutes it existed out of its native element, had predicted a certain number of years of wonderful fertility and a fatal epidemic, the only remedy for which would be the possession of the marine prophet's likeness! The sale of these *pictured Mermaids* was immense. Either the composite animal, or another, the offspring of the success of the first, was sold to the Dutch factory and transmitted to Batavia, where it fell into the hands of a speculating American, who brought it to Europe; and here, in the year 1822-3, exhibited his purchase as a real Mermaid to the admiration of the ignorant, the perplexity of the learned, and the filling of his own purse."

The Editor of the "Literary Gazette," Mr. Jerdan, was the first to expose the fabulous creature of the Egyptian Hall. He plainly said:—"Our opinion is fixed that it is a *composition;* a most ingenious one, we grant, but still nothing beyond the admirably put-together members of various animals. The extraordinary skill of the Chinese and Japanese in executing such deceptions is notorious, and we have no doubt that the Mermaid is a manufacture from the Indian Sea, where it has been pretended it was caught. We are not of those who because they happen not to have had direct proof of the existence of any extraordinary natural phenomenon, push scepticism to the extreme and deny its possibility. The depths of the sea, in all probability, from various chemical and philosophical causes, contain animals unknown to its surface-waters, rarely if ever seen by human eye. But when a creature is presented to us having no other organization but that which is suitable to a medium always open to our observation, it in the first instance excites suspicion that only one individual of the species should be discovered and obtained. When knowledge was more limited, the stories of Mermaids seen in distant quarters might be credited by the many, and not entirely disbelieved by the few; but now, when European and especially British commerce fills every corner of the earth with men of observation and science, the unique becomes the incredible, and we receive with far greater doubt the apparition of such anomalies as the present. It is curious that though medical men seem in general to regard the creature as a possible production of nature, no naturalist of any ability credits it after five minutes' observation! This may, perhaps, be accounted for by their acquaintance with the parts of distinct animals, of which it appears the Mermaid is composed. The cheeks of the blue-faced ape, the canine teeth, the *simia* upper body, and the tail of the fish, are all familiar to them in less complex combinations, and they pronounce at once that the whole is an imposture. And such is our settled conviction." Though naturalists and journalists fully exposed the imposture, this did not affect the exhibition, which for a considerable time continued as crowded as ever; but the notoriety had dwindled down to "a penny show," at Bartholomew Fair, by the year 1825.

After so many exposures of the absurd belief in Mermaids, it could scarcely be expected that any person could be found in Europe weak enough to report the existence of one of these creatures to an eminent scientific Society. Yet, on the 22d of June, 1840, the first Secretary of the Ottoman Embassy at Paris addressed a note to the Academy of Sciences, stating that his father, who was in the Admiralty department at Constantinople, had recently seen a Mermaid while crossing the Bosphorus, which communication was received with much laughter.

We have still another recorded instance—and in Scotland. In the year 1857 two fishermen on the Argyleshire coast declared that when on their way to the fishing-station, Lochindale, in a boat, and when about four miles south-west from the village of Port Charlotte, about six o'clock in a June evening, they distinctly saw, at about six yards distance, an object in the form of a woman, with comely face and fine hair hanging in ringlets over the neck and shoulders. It was above the surface of the water gazing at the fishermen for three or four minutes

—and then vanished! Yet this declaration was officially attested!

In 1863 Mermaids were supposed to abound in the ponds and ditches of Suffolk, where careful mothers used them as bugbears to prevent little children from going too near the water. Children described them as "nasty things that crome you (hook you) into the water;" others as "a great big thing like a feesh," probably a pike basking in the shallow water.

Sometimes the Mermaid has assumed a picturesqueness in fairy tale; and her impersonation has been described by Dryden as "a fine woman, with a fish's tail." And, laying aside her scaly train, she has appeared as a lovely woman, with sea-green hair; and Crofton Croker relates, in his "Fairy Legends," a marriage between an Irish fisherman and a "Merrow," as the Mermaid is called in Ireland.

IS THE UNICORN FABULOUS?

TO this question we may reply, in the words of a writer of 1633, "Concerning the Unicorn, different opinions prevail among authors: some doubt, others deny, and a third class affirm its existence." The question has lasted two thousand years, and is every now and then kept alive by fresh evidences.

Ctesias, a credulous Greek physician, who appears to have resided at the Court of Persia, in the time of the younger Cyrus, about 400 years before the birth of Christ, describes the wild asses of India as equal to the horse in size, and even larger, with white bodies, red heads, bluish eyes, and a horn on the forehead a cubit in length; the part from the forehead entirely white, the middle black, and the extremity red and pointed. Drinking-vessels were made of it, and those who used them were subject neither to convulsions, epilepsy, nor poison, provided that before taking the poison, or after, they drank from these cups water, wine, or any other liquor. Ctesias describes these animals as very swift and very strong. Naturally they were not ferocious; but when they found themselves and their young surrounded by horsemen, they did not abandon their offspring, but defended themselves by striking with their horns, kicking, and biting, and so slew many men and horses. This animal was also shot with arrows and brought down with darts; for it was impossible to take it alive. Its flesh was too bitter for food, but it was hunted for its horn and astragalus (ankle-bone), which last Ctesias declares he saw. Aristotle describes the Indian ass with a single horn. Herodotus mentions asses having horns; and Strabo refers to Unicorn horses, with the heads of deers. Oppian notices the Aonian bulls with undivided hoofs, and a single median horn between their temples. Pliny notices it as a very ferocious beast, similar in its body to a horse, with the head of a deer, the feet of an elephant, the tail of a boar, a deep bellowing voice, and a single black horn standing out in the middle of its forehead. He adds, that it cannot be taken alive; and some such excuse may have been necessary in those days for not producing the living animal upon the arena of the amphitheatre.

Out of this passage most of the modern Unicorns have been described and figured. The body of the horse and the head of the deer appear to be but vague sketches; the feet of the elephant and the tail of the boar point at once to a pachydermatous (thick-skinned) animal; and the single black horn, allowing for a little exaggeration as to its length, well fits the two last-mentioned conditions, and will apply to the Indian rhinoceros, which, says the sound naturalist, Ogilby, "affords a remarkable instance of the obstructions which the progress of knowledge may suffer, and the gross absurdities which not unfrequently result from the wrong application of a name." Mr. Ogilby then refers to the account of Ctesias, which we have just quoted, and adds :—"His account, though mixed up with a great deal of credulous absurdity, contains a very valuable and perfectly recognisable description of the rhinoceros, under the ridiculous name, however, of the *Indian Ass;* and, as he attributed to it a whole hoof like the horse, and a single horn in the forehead, speculation required but one step further to produce the fabulous Unicorn."

The ancient writers who have treated of the Unicorn are too numerous for us to specify. Some of the moderns may be referred to. Garcias describes this marvellous creature from one who alleges that he had seen it. The seer affirmed that it was endowed with a wonderful horn, which it would sometimes turn to the left and right, at others raise, and then again depress. Ludovicus Vartomanus writes, that he saw two sent to the Sultan from Ethiopia, and kept in a repository at Mahomet's tomb in Mecca. Cardan describes the Unicorn as a rare animal, the size of a horse, with hair very like that of a weasel, with the head of a deer, on which one horn grows three cubits in length (a story seldom loses anything in its progress) from the forehead, ample at its lowest part, and tapering to a point; with a short neck, a very thin mane, leaning to one side only, and less on the ear, as those of a young roe.

In Jonston's "Historia Naturalis," 1657, we see the smooth-horned solipede, "Wald Esel;" and the digitated and clawed smooth-horned "Meer Wolff," the latter with his single horn erect in the foreground, but with it depressed in the background, where he is represented regaling on serpents. Then there are varieties, with the head, mane, and tail of a horse; another smooth-horned, with a horse's head and mane, a pig's-tail and camel-like feet; the "Meer Stenbock, Capricornus Marinus," with hind webbed feet, and a kind of graduated horn, like an opera-glass pulled out, in the foreground, and charging the fish most valiantly in the water in the distance. Then there is another, with a mule's head and two rhinoceros-like horns, one on his forehead and the other on his nose; and a horse's tail, with a collar round his neck; a neck entirely shaggy—and a twisted horn, a shaggy gorget, and curly tail, are among other peculiarities.

The Unicorn seems to have been a sad trouble to the hunters, who hardly knew how to come at so valuable a piece of game. Some described the horn as moveable at the will of the animal—a kind of small sword, in short, with which no hunter who was not exceedingly cunning in fence could have a chance. Others told the poor foresters that all the strength lay in its horn, and that when pressed by them it would throw itself from the pinnacle of the highest rock, horn foremost, so as to pitch upon it, and then quietly march off not a bit the worse!

Modern zoologists, disgusted as they well may be with fables, such as we have glanced at, disbelieve, generally, the existence of the Unicorn, such, at least, as we have referred to; but there is still an opinion that some land animal bearing a horn on the anterior part of its head, exists besides the rhinoceros. The nearest approach to a horn in the middle of the forehead of any terrestrial mammiferous animal known to us is the bony protuberance on the forehead of the giraffe; and though it would be presumptuous to deny the existence of a one-horned quadruped other than the rhinoceros, it may be safely stated that the insertion of a long and solid horn in the living forehead of a horse-like or deer-like cranium is as near an impossibility as anything can be.

Rupell, after a long sojourn in the north-east of Africa, stated that in Kordofan the Unicorn exists; stated to be the size of a small horse, of the slender make of the gazelle, and furnished with a long straight horn in the male, which was wanting in the female. According to the statements made by various persons, it inhabits the deserts to the south of Koretofan, is uncommonly fleet, and comes only occasionally to the Koldagi Heive mountains on the borders of Kordofan.

Other writers refer the Unicorn to the antelope. The origin of the name of antelope is traced by

Cuvier to the Greek *Anthalops,* applied to a fabulous animal living on the banks of the Euphrates, with long jagged horns, with which it sawed down trees of considerable thickness! Others conjecture this animal to have been the *Oryx,* a species of antelope, which is fabulously reported to have had only one horn, and to have been termed *Panthalops* in the old language of Egypt.

In his "Revolutions on the Surface of the Globe," Cuvier refers the idea of the Unicorn to the coarse figures traced by savages on rocks. Ignorant of perspective, and wishing to present in profile the horned antelope, they could only give it one horn; and thus originated the *Oryx.* The oryx of the Egyptian monuments is, most probably, but the production of a similarly crude style, which the religion of the country imposed on the artist. Many of the profiles of quadrupeds have only one leg before and one behind: why, then, should they show two horns? It is possible that individual animals might be taken in the chase whom accident had despoiled of one horn, as it often happens to chamois and the Scythian antelope; and that would suffice to confirm the error which these pictures originally produced. It is thus, probably, that we find anew the Unicorn in the mountains of Thibet.

The *Chiru Antelope* is the supposed Unicorn of the Bhotians. In form it approaches the deer; the horns are exceedingly long, are placed very forward in the head, and may be popularly described as erect and straight. It is usually found in herds, and is extremely wild, and unapproachable by man. It is much addicted to salt in summer, when vast herds are often seen at the rock-salt beds which abound in Tibet. They are said to advance under the conduct of a leader, and to post sentinels around the beds before they attempt to feed.

Major Salter is stated to have obtained information of the existence of an animal in Tibet closely resembling the Unicorn of the ancients, which revived the belief of naturalists by adducing testimonies from Oriental writings. Upon this statement, M. Klaproth remarks, that previous to Major Salter's Reports, the Catholic missionaries, who returned to Europe from China by way of Tibet and Nepal, in the seventeenth century, mentioned that the Unicorn was found in that part of the Great Desert which bounds China to the west, where they crossed the great wall; that Captain Turner, when travelling in Tibet, was informed by the Raja of Boutan that he had one of these animals alive; and that Bell, in his "Travels to Peking," describes a Unicorn which was found on the southern front of Siberia. He adds:—"The great 'Tibetan-Mongol Dictionary' mentions the Unicorn; and the 'Geographical Dictionary of Tibet and Central Asia,' printed at Peking, where it describes a district in the province of Kham, in Tibet, named Sera-zeong, explains this name by 'the River of Unicorns,' because," adds the author, "many of these animals are found there."

In the "History of the Mongol-Khans," published and translated at St. Petersburg, we find the following statement:—Genghiz Khan, having subjected all Tibet in 1206, commenced his march for Hindustan. As he ascended Mount Jadanarung, he beheld a beast approaching him of the deer kind, of the species called *Seron*, which have a single horn at the top of the head. It fell on its knees thrice before the monarch, as if to pay respect to him. Every one was astonished at this incident. The monarch exclaimed, "The Empire of Hindustan is, we are assured, the country where are born the majestic Buddhas and Bodhisatwas, as well as the potent Bogdas and princes of antiquity: what can be the meaning, then, of this animal, incapable of speech, saluting me like a man?" Upon this, he returned to his own country. "This story," continues M. Klaproth, "is also related by Mahommedan authors who have written the life of Genghiz. Something of the kind must, therefore, have taken place. Possibly, some of the Mongol conqueror's suite may have taken a Unicorn, which Genghiz thus employed, to gain a pretext for abstaining from an expedition which promised no success."

Upon this statement, it was observed in the "Asiatic Register," 1839, that "when we consider that seventeen years have elapsed since the account of Major Salter was given, and that, notwithstanding our increased opportunities of intercourse with Tibet, no fact has since transpired which supplies a confirmation of that account, except the obtaining of a supposed horn of the supposed Unicorn, we cannot participate in these renewed hopes."

The Rev. John Campbell, in his "Travels in South Africa," describes the head of another animal, which, as far as the horn is concerned, seems to approach nearer than the common rhinoceros to the Unicorn of the ancients. While in the Machow territory, the Hottentots brought to Mr. Campbell a head differing from that of any rhinoceros that had been previously killed. "The common African rhinoceros has a crooked horn, resembling a cock's spur, which rises about nine or ten inches above the nose, and inclines backward; immediately behind which is a straight thick horn. But the head brought by the Hottentots had a straight horn projecting three feet from the forehead, about ten inches above the tip of the nose. The projection of this great horn very much resembles that of the fanciful Unicorn in the British arms. It has a small thick horny substance, eight inches long, immediately behind it, which can hardly be observed on the animal at the distance of a hundred yards; so that this species must look like an Unicorn (in the sense 'one-horned') when running in the field." The author adds:—"This animal is considered by naturalists, since the arrival of the above skull in London, to be the Unicorn of the ancients, and the same that is described in Job xxxix. 9—'Will the Unicorn be willing to serve thee, or abide by thy crib? 10. Canst thou bind the Unicorn with his band in the furrow? or will he harrow the valleys after thee? 11. Wilt thou trust him because his strength is great? or wilt thou leave thy labour to him?' Again, Deuteronomy xxxiii. 17—

'His horns are like the horns of Unicorns: with them he shall push the people together to the ends of the earth.'"

A fragment of the skull, with the horn, is deposited in the Museum of the London Missionary Society.

Mr. W. B. Baikie writes to the *Athenæum* from Bida Núpe, Central Africa, in 1862, the following suggestions:—"When I ascended the Niger, now nearly five years ago, I frequently heard allusions to an animal of this nature, but at that time I set it down as a myth. Since then, however, the amount of testimony I have received, and the universal belief of the natives of all the countries which I have hitherto visited, have partly shaken my scepticism, and at present I simply hold that its non-existence is not proven. A skull of this animal is said to be preserved in a town in the country of Bonú, through which I hope to pass in the course of a few weeks, when I shall make every possible inquiry. Two among my informants have repeatedly declared to me that they have seen the bones of this animal, and each made particular mention of the long, straight, or nearly straight, black horn. In countries to the east, and south-east, as Márgi and Bagirmi, where the one-horned rhinoceros is found, the hunters carefully distinguished between it and the supposed Unicorn, and give them different names. In the vast forests and boundless wastes which occur over Central Africa, especially towards the countries south and east of Lake Tsád, Bórnú, Bagirmi and

Adamáwa, are doubtless numerous zoological curiosities as yet unknown to the man of science, and among them possibly may exist this much-talked-of, strange, one-horned animal, even though it may not exactly correspond with our typical English Unicorn."

The factitious horn has been preserved in various Museums. The " Monocero Horn," in Tradescant's collection, was, probably, that which ordinarily has passed for the horn of the Unicorn, namely, the tooth of a narwhal. Old legends assert that the Unicorn, when he goes to drink, first dips his horn in the water to purify it, and that other beasts delay to quench their thirst till the Unicorn has thus sweetened the water. The narwhal's tooth makes a capital twisted Unicorn's horn, as represented in the old figures. That in the Repository of St. Denis, at Paris, was presented by Thevet, and was declared to have been given to him by the King of Monomotapa, who took him out to hunt Unicorns, which are frequent in that country. Some have thought that this horn was a carved elephant's tooth. There is one at Strasburg, some seven or eight feet in length, and there are several in Venice.

Great medical virtues were attributed to the so-called horn, and the price it once bore outdoes everything in the *Tulipomania*. A Florentine physician has recorded that a pound of it (sixteen ounces) was sold in the shops for fifteen hundred and thirty-six crowns, when the same weight in gold would only have brought one hundred and forty-eight crowns.

From what source we derive the stories of the animosity between the lion and the Unicorn is not clearly understood, although this is the principal medium through which the fabulous creature has been kept in remembrance by being constantly before us in the Royal Arms, which were settled at the Accession of George I. We owe the introduction of the Unicorn, however, to James I., who, as King of Scotland, bore two Unicorns, and coupled one with the English lion, when the two kingdoms were united.

The position of the lion and Unicorn in the arms of our country seems to have given rise (naturally enough in the mind of one who was ignorant of heraldic decoration) to a nursery rhyme which most of us remember:—

> "The Lion and the Unicorn
> Were fighting for the crown;
> The Lion beat the Unicorn
> All round the town," &c.

unless it alludes to a contest for dominion over the brute creation, which the "rebellious Unicorn," as Spenser calls it, seems to have waged with the tawny monarch.

Spenser, in his "Faerie Queen," gives the following curious way of catching the Unicorn:—

> "Like as a lyon, whose imperiall powre,
> A prowd rebellious Unicorn defyes,
> T'avoide the rash assault and wrathful stowre
> Of his fiers foe, him a tree applyes,
> And when him rousing in full course he spyes,

> He slips aside; the whiles that furious beast
> His precious horne, sought of his enemyes,
> Strikes in the stocke, ne thence can be releast,
> But to the mighty victor yields a bounteous feast."

Shakspeare, also ("Julius Cæsar," Act ii. scene 1), speaks of the supposed mode of entrapping them:—

> "For he loves to hear
> That Unicorns may be betrayed with trees,
> And bears with glasses, elephants with holes,
> Lions with toils, and men with flatterers."

We have no satisfactory reason for believing that man ever coexisted with Mastodons; otherwise Professor Owen's discovery of the retention of a single tusk only by the male gigantic Mastodon, might have afforded another form of Unicorn.

Whatever the zoologists may have done towards extirpating the belief in the existence of the Unicorn, it is ever kept in sight by heraldry, which, with its animal absurdities, has contributed more to the propagation of error respecting the natural world than any other species of misrepresentation.

THE MOLE AT HOME.

THE Mole, though generally a despised and persecuted animal, is nevertheless useful to the husbandman in being the natural drainer of his land and destroyer of worms. To other inferior animals he is a sapper and miner, forming for them their safe retreats and well-secured dormitories.

The economy of the Mole has been much controverted among naturalists. It is found throughout the greater part of Europe. We are overrun with it in most parts of England and Wales; but it does not appear to have been found in the northern extremity of Scotland, and there is no record of its having been seen in the Orkney Isles, Zetland, or Ireland. Its most diligent and instructive historian is Henri Le Court, who, flying from the terrors that came in the train of the French Revolution, betook himself to the country, and from being the attendant on a Court, became the biographer of this humble animal. M. Geoffroy St. Hilaire, the celebrated French naturalist, visited Le Court for the purpose of testing his observations, and appears to have been charmed by the facility and ingenuity with which Le Court traced and demonstrated the subterraneous labours of this obscure worker in the dark.

We shall first briefly describe the adaptation of its structure to its habits. The bony framework is set in motion by very powerful muscles, those of the chest and neck being most vigorous. The wide hand, which is the great instrument of action, and performs the offices of a pickaxe and shovel, is sharp-edged on its lower margin, and when clothed with the integuments the fingers are hardly distinguishable. The muzzle of the Mole is evidently a delicate organ of touch, as are also the large and broad hands and feet; and the tail has much sensitiveness to give notice to the animal of the approach of any attack from behind. Its taste and smell, especially the latter, are very sensitive. Its sight is almost rudimentary. The little eye is so hidden in the fur that its very existence was for a long time doubted. It appears to be designed for operating only as a warning to the animal on its emerging into the light; indeed, more acute vision would only have been an encumbrance. If the sight be imperfect, the sense of hearing is very acute, and the tympanum very large, though there is no external ear, perhaps because the earth assists considerably in vibration. The fore-feet are inclined sideways, so as to answer the use of hands, to scoop out the earth to form its habitation or pursue its prey, and to fling all the loose soil behind the animal. The breastbone in shape resembles a ploughshare. The skin is so tough as only to be cut by a very sharp knife. The hair is very short and close-set, and softer than the finest silk; colour black; some spotted and cream-coloured. This hair is yielding; had it been strong, as in the rat or mouse, it would doubly have retarded the progress of the creature; first by its resistance, and then acting as a brush, so as to choke up the galleries, by removing the loose earth from the sides and ceilings of the galleries.

It is supposed that the verdant circles so often seen in grass ground, called by country people *fairy rings*, are owing to the operations of Moles: at certain seasons they perform their burrowings in circles, which, loosening the soil, gives the surface a greater fertility and rankness of grass than the other parts within or without the ring. The larger mole-hills denote the nests or dens of the Mole beneath.

The feeling of the Mole is so acute that when casting up the earth, it is sensible of very gentle pressure; hence molecatchers tread lightly when in quest of Moles; and unless this caution is used the Mole ceases its operation, and instantly retires. Again, so acute is the smell, that molecatchers draw the body of a captured Mole through their traps and the adjoining runs and passages to remove all suspicious odours which might arise from the touch of their fingers.

During summer the Mole runs in search of snails

and worms in the night-time among the grass, which pursuit makes it the prey of owls. The Mole shows great art in skinning a worm, which it always does before it eats it, by stripping the skin from end to end, and squeezing out the contents of the body. It is doubtful whether any other animal exists which is obliged to eat at such short intervals as the Mole, ten or twelve hours appearing to be the maximum of its fasting; at the end of that time it dies. Cuvier tells us that if two Moles are shut up together without food, there will shortly be nothing left of the weakest but its skin, slit along the belly! Buffon accuses Moles of eating all the acorns of a newly-set soil. Its voracity makes the Mole a great drinker: a run is always formed to a pond or ditch as a reservoir; when it is too distant, the animal sinks little wells, which have sometimes been seen brimfull.

We now return to Le Court's experiments with Moles, which are very interesting. To afford proof of the rapidity with which the Mole will travel along its passages, Le Court watched his opportunity, and when the animal was on its feed at one of the most distant points from its sanctuary or fortress, to which point the Mole's high road leads, Le Court placed along the course of that road, between the animal and the fortress, several little camp colours, so to speak, the staff of each being a straw, and the flag a bit of paper, at certain distances, the straws penetrating down into the passage. Near the end of this subterraneous road he inserted a horn, the mouthpiece of which stood out of the ground. When all was ready, Le Court blew a blast loud enough to frighten all the Moles within hearing. Down went the little flags in succession with astonishing velocity, as the terrified Mole, rushing along towards his sanctuary, came in contact with the flag-straws; and the spectators affirmed that the Mole's swiftness was equal to the speed of a horse at a good round trot.

To test its amount of vision, Le Court took a spare water-pipe, or gutter, open at both ends. Into this pipe he introduced several Moles successively. Geoffroy St. Hilaire stood by to watch the result at the further end of the tube. As long as the spectators stood motionless the introduced Mole made the best of his way through the pipe and escaped; but if they moved, or even raised a finger, the Mole stopped, and then retreated. Several repetitions of this experiment produced the same results.

In the domain of the Mole, the principal point is the habitation, or fortress, constructed under a considerable hillock raised in some secure place, often at the root of a tree, or under a bank. The dome of the fortress is of earth, beaten by the Mole-architect into a compact and solid state. Inside is formed a circular gallery at the base, which communicates with a smaller upper gallery by means of five passages. Within the lower gallery is the chamber or dormitory, which has access to the upper gallery by three passages. From this habitation extends the high road by which the proprietor reaches the opposite end of the encampment; the galleries open into this road, which the Mole is continually carrying out and extending in his search for food; this has been termed the *hunting-ground*. Another road extends, first downwards, and then up into the open road of the territory. Some eight or nine other passages open out from the external circular gallery. From the habitation a road is carried out, nearly straight, and connected with the encampment and the alleys leading to the hunting-ground which open into it on each side. In diameter the road exceeds the body of a Mole, but its size will not admit of two Moles passing each other. The walls, from the repeated pressure of the animal's sides, become smooth and compact. Sometimes a Mole will lay out a second or even a third road; or several individuals use one road in common, though they never trespass on each other's hunting-grounds.

If two Moles should happen to meet in the same road, one must retreat into the nearest alley, unless they fight, when the weakest is often slain. In forming this tunnel the Mole's instinct drives it at a greater or less depth, according to the quality of the soil, or other circumstances. When it is carried under a road or stream, a foot and a-half of earth, or sometimes more, is left above it. Then does the little engineering Mole carry on the subterraneous works necessary for his support, travelling, and comfort; and his tunnels never fall in. The quality or humidity of the soils which regulates the abundance of earth-worms, determines the greater or less depth of the alleys; and when these are filled with stores of food the Mole works out branch alleys.

The main road communicating with the hunting-grounds is of necessity passed through in the course of the day; and here the mole-catcher sets his traps to intercept the Mole between his habitation and the alley where he is carrying on his labours. Some mole-catchers will tell you the hours when the Moles move are nine and four; others that near the coast their movements are influenced by the tides. Besides the various traps which are set for Moles, they are sometimes taken by a man and a dog; when the latter indicates the presence of a Mole, the man spears the animal out as it moves in its run. Pointers will stop as steadily as at game, at the Moles, when they are straying on the surface.

The Mole is a most voracious animal. Earthworms and the larvæ of insects are its favourite food; and it will eat mice, lizards, frogs, and even birds; but it rejects toads, even when pressed by hunger, deterred, probably, by the acrid secretions of their skin. Moles are essentially carnivorous; and when fed abundantly on vegetable substances they have died of hunger.

During the season of love, at which time fierce battles are fought between the males, the male pursues the female with ardour through numerous runs wrought out with great rapidity. The attachment appears to be very strong in the Moles. Le Court often found a female taken in his trap and a male lying dead close to her. From four to five is the general number of young. The nest is distinct, usually distant from the habitation. It is constructed by enlarging and excavating the point where three or four passages intersect each other; and the bed of the nest is formed of a mass of young grass, root fibres, and herbage. In one nest Geoffroy St. Hilaire and Le Court counted two hundred and four young wheat-blades.

M. St. Hilaire describes the pairings, or as he calls it, "the loves of the Moles." As soon as the Mole has finished the galleries he brings his mate along with him, and shuts her up in the bridal gallery, taking care to prevent the entrance of his rivals: in case of a fight they enlarge the part of the gallery where they are met; and the victory is decided in favour of him who first wounds his adversary before the ear. The female, during the fight, is shut up in the bridal gallery, so as to be unable to escape; for which purpose, however, she uses all her resources in digging, and attempts to get away by the side passages. Should she succeed the conqueror hastens to rejoin his faithless mate, and to bring her back into his galleries. This manœuvre is repeated as often as other males enter the lists. At length the conqueror is recognised, and his mate becomes more docile. The pair work together and finish the galleries; after which the female digs alone for food. As soon as the galleries are formed, the male conducts his mate to a certain point, and from this time the female no longer digs in the solid earth, but towards the surface, advancing by merely separating the roots of the grass.

The Mole is a great friend to the farmer; but there are places in which he is a public enemy. He is not a vegetable feeder, and he never roots up the growing corn in spring-time, except when he is after grubs, snails, and wire-worms. It has been calculated that two Moles destroy 20,000 white worms in a year. He is very destructive to under drains; and where the land is low we are in danger of a deluge from his piercing holes in the drain-banks. Thus it would be madness not to extirpate Moles in those places where the waters, in drains or rivers, are above the level of the lands around, especially when the banks are made of sand or earth of loose texture.

The persecution of Moles in cultivated countries amounts almost to a war of extermination. The numbers annually slaughtered are enormous. A mole-catcher, who had followed the craft for thirty-five years, destroyed from forty to fifty thousand Moles. But all Mole exterminators must yield to Le Court, who, in no large district, took, in five months, six thousand of them. Moles are good swimmers, and their bite is very sharp; their attacks are ferocious, and they keep their hold like a bull-dog.

The Shrew Mole of North America resembles the European Mole in its habits. Dr. Goodman describes it as most active early in the morning, at mid-day, and in the evening; and they are well known in the country to have the custom of coming daily to the surface *exactly at noon*. We read of a captive Shrew Mole which ate meat, cooked or raw, drank freely, and was lively and playful, following the hand of his feeder by the scent, burrowing for a short distance in the loose earth, and after making a small circle, returning for more food. In eating he employed his flexible snout to thrust the food into his mouth, doubling it so as to force it directly backwards, as described in Dr. Richardson's "North American Zoology."

James Hogg, the Ettrick Shepherd, remarks, in his usual impressive manner :—"The most unnatural persecution that ever was raised in a country is that against the Mole—that innocent and blessed little pioneer, who enriches our pastures annually with the first top-dressing, dug with great pains and labour from the fattest of the soil beneath. The advantages of this top-dressing are so apparent that it is really amazing how our countrymen should have persisted, for nearly half a century, in the most manly and valiant endeavours to exterminate the Moles! If a hundred men and horses were employed on a pasture farm of from fifteen hundred to two thousand acres, in raising and driving manure for a top-dressing of that farm, they would not do it so effectually, so neatly, or so equally as the natural number of Moles. In June, July, and August, the Mole-hills are all spread by the crows and lambs—the former for food, and the latter in the evenings of warm days after a drought has set in. The late Duke of Buccleuch was the first who introduced Mole-catching into Scotland."

THE GREAT ANT-BEAR.

A FINE living specimen of this comparatively rare animal was first exhibited in the Zoological Society's gardens, in the Regent's-park, 1853. It is stated to be the first specimen brought alive to England, and accordingly excited considerable attention. It was one of a pair, captured near the Rio Negro, in the southern province of Brazil, and shipped for England by some German travellers. The male died on the voyage; the female arrived in London in 1853, and was exhibited in Broad-street, St. Giles's, until purchased by the Zoological Society for the sum of 200*l*. The advantage of this live specimen to naturalists has been very great. Hitherto the examples engraved by Buffon and Shaw were both derived from stuffed specimens, and had the inevitable defects and shortcomings of such. Sir John Talbot Dillon, in his "Travels through Spain," published in 1780, states that a specimen of the Ant-Bear, from Buenos Ayres, was alive at Madrid in 1776: it is now stuffed and preserved in the Royal Cabinet of Natural History at Madrid. The persons who brought it from Buenos Ayres say it differs from the Ant-eater, which only feeds on emmets and other insects, whereas this would eat flesh, when cut in small pieces, to the amount of four or five pounds. From the snout to the extremity of the tail this animal is two yards in length, and his height is about two feet; the head very narrow, the nose long and slender. The tongue is so singular that it looks like a worm, and extends above sixteen inches. The body is covered with long hair of a dark brown, with white stripes on the shoulders; and when he sleeps, he covers his body with his tail. This account, it will be seen hereafter, corresponds very accurately with that of the animal purchased by the Zoological Society.

Mr. Wallace, who travelled on the Amazon and Rio Negro, about the year 1853, relates:—" The living specimen of this singular animal is a great rarity, even in its native country. In fact, there is not a city in Brazil where it would not be considered almost as much a curiosity as it is here. In the extensive forests of the Amazon the great Ant-eater is, perhaps, as abundant as in any part of South America; yet, during a residence there of more than four years, I never had an opportunity of seeing one. Once only I was nearly in at the death, finding a bunch of hairs from the tail of a specimen which had been killed (and eaten) a month previous to my arrival, at a village near the Capiquiare. In its native forests the creature feeds almost entirely on white ants, tearing open their nests with its powerful claws, and thrusting in its long and slender tongue, which, being probably mistaken for a worm, is immediately seized by scores of the inhabitants, who thus become an easy prey. The Indians, who also eat white ants, catch them in a somewhat similar manner, by pushing into the nest a grass-stalk, which the insects seize and hold on to most tenaciously. It may easily be conceived that such an animal must range over a considerable extent of country to obtain a plentiful supply of such food, which circumstance, as well as its extreme shyness and timidity, causes it to be but rarely met with, and still more rarely obtained alive."

We have seen that the Ant-Bear lives exclusively upon ants, to procure which he tears open the hills, and when the ants flock out to defend their dwellings, draws over them his long, flexible tongue, covered with glutinous saliva, to which the ants consequently adhere; and he is said to repeat this operation twice in a second. "It seems almost incredible," says Azara, "that so robust and powerful an animal can procure sufficient sustenance from ants alone; but this circumstance has nothing strange in it, for those who are acquainted with the tropical parts of America, and have seen the enormous multitude of these insects, which swarm in all parts of the country to that degree that their hills often almost touch one another for miles together." The same author informs us that domestic Ant-Bears were occasionally kept by different persons in Paraguay, and that they had even been sent alive to Spain, being fed upon bread-and-milk mixed with morsels of flesh minced very small. Like all animals which live upon insects, the Ant-eaters are capable of sustaining a total deprivation of nourishment for an almost incredible time.

The Great Ant-Bear's favourite resorts are low, swampy savannahs, along the banks of rivers and stagnant ponds; also frequenting humid forests, but never climbing trees, as falsely reported by Buffon. His pace is slow and heavy, though, when hard pressed, he increases his rate, yet his greatest velocity never half equals the ordinary running of a man. When pressed too hard, or urged to extremity, he turns obstinate, sits upon his hind-quarters like a bear, and defends himself with his powerful claws. Like that animal, his usual and only mode of assault is by seizing his adversary with his fore-paws, wrapping his arms round him, and endeavouring, by this means, to squeeze him to death. His great strength and powerful muscles would easily enable him to accomplish his purpose in this respect, even against the largest animals of his native forests, were it but guided by ordinary intelligence, or accompanied with a common degree of activity; but in these qualities there are few animals indeed who do not greatly surpass the Ant-Bear; so that the different stories handed down by writers on natural history from one to another, and copied, without question, into the histories and descriptions of this animal, may be regarded as pure fictions. "It is supposed," says Don Felix d'Azara, "that the jaguar himself dares not attack the Ant-Bear, and that, if pressed by hunger, or under some other strong excitement, he does so, the Ant-Bear embraces and hugs him so tightly as very soon to deprive him of life, not even relaxing his hold for hours after life has been extinguished in his assailant. Such is the manner in which the Ant-eater defends himself; but it is not to be believed that his utmost efforts could prevail against the jaguar, who, by a single bite, or blow of his paw, could kill the Ant-eater before he was prepared for resistance, so slow are his motions, even in an extreme case; and, being unable to leap or turn with ordinary rapidity, he is forced to act solely upon the defensive. The flesh of the Ant-eater is esteemed a delicacy by the Indians; and, though black, and of a strong musky flavour, is sometimes even met with at the tables of Europeans."

The habits of the Great Ant-Bear in captivity have been described scientifically yet popularly, from the Zoological Society's specimen, by Professor Owen, who writes:—"When we were introduced to this, the latest novelty at the noble vivarium in the Regent's-park, we found the animal busy sucking and licking up—for his feeding is a combination of the two actions—the contents of a basin of squashed eggs. The singularly long and slender head, which looks more like a slightly bent proboscis, or some

such appendage to a head, was buried in the basin, and the end of the lithe or flexible tongue, like a rat's tail, or a writhing black worm, was ever and anon seen coiling up the sides of the basin, as it was rapidly protruded and withdrawn. The yellow yolk was dripping with the abundant ropy saliva secreted during the feeding process from the exceedingly small terminal mouth; for the jaws are not slit open, as in the ordinary construction of the mouths of quadrupeds, and the head, viewed sideways, seems devoid of mouth; but this important aperture—by some deemed the essential character of an animal—is a small orifice or slit at the end of the tubular muzzle, just being enough, apparently, to let the vermiform tongue slip easily in and out. The tongue, the keeper told us, was sometimes protruded as far as fourteen inches from the mouth."

By the Qjuarani Indians the beast is known by a name which is, in Spanish, "little mouth." The Portuguese and Spanish peons call it by a name equivalent to "Ant-Bear." In the Zoological Catalogue the animal is denominated *Myrmocophaga jubata*, or the "Maned Ant-eater." This appellation would very well suit the animal if, as most spectators commonly imagine at first sight, its head was where its tail is, for the tail is that part of the animal on which the hair is most developed, after the fashion of a mane; whilst the actual head appears much more like a tail, of a slender, almost naked, stiff, rounded kind. The body is wholly covered by long, coarse hair, resembling hay, rapidly lengthening from the neck backwards to six or eight inches, and extending on the tail from ten to eighteen inches. The colour is greyish brown, with an oblique black band, bordered with white, on each shoulder. The animal measures about four feet from the snout to the root of the tail; and the tail, three feet long, resembles a large screen of coarse hair. When the animal lies down, it bends its head between its fore legs, slides these forward, and crosses them in front of the occiput, sinks its haunches by bending its hind legs and bringing them close to the fore feet; then, leaning against the wall of its den, on one side, it lays the broad tail over the other exposed side of the body, by the side bend of that part, like the movement of a door or screen. Nothing is now visible of the animal but the long coarse hair of its *natural and portable blanket*. When it is enjoying its siesta, you cannot form any conception of its very peculiar shape and proportions; an oblong heap of a coarse, dry, *greyish thatch* is all that is visible. When, however, the keeper enters the den with any new dainty, as cockroaches, crickets, maggots, or mealworms, to tempt the huge insect-devourer, the quick-hearing animal unveils its form by a sweeping movement of the thatch outwards, the tail that supports it rotating, as if joined by a kind of door-hinge to the body; the head is drawn out from between the fore limbs; the limbs are extended, and the entire figure of this most grotesque of quadrupeds stalks forth. The limbs are short; the fore limbs grow rather thicker to their stumpy ends, which look as if the feet had been amputated. The four toes, with their claws, are bent inwards, and are of very unequal length. This is the most singular part of the animal: it is also the most formidable member, and, indeed, bears the sole weapon of defence the beast possesses. The innermost toe, answering to the thumb on the fore limb of the neighbouring chimpanzee, is the shortest. A fifth toe seems to be buried in the outside callosity, on which the animal rests its stumpy feet while walking. At the back part of the sole, or palm, of the fore foot, is a second large callosity, which receives the point of the great claw in its usual state of inward inflection. Against this callosity the animal presses the claw when it seizes any object therewith; and Azara, as we have seen, avers that nothing can make the Ant-Bear relax its grasp of an object so seized.

With respect to the jaguar being sometimes found dead in the grasp of the Great Ant-eater, Professor Owen observes that its muscular force resembles that of the cold-blooded reptiles in the force and endurance of the contractile action; and, like the reptiles, the Sloths and Ant-Bears can endure long fasts.

Woe to the unlucky or heedless aggressor whose arm or leg may be seized by the Ant-Bear. The strength of the grasp sometimes breaks the bone. The Ant-Bear never voluntarily lets go, and the limb so grasped can be with difficulty extricated, even after the animal has been killed. To put the beast, however, *hors de combat*, no other weapon is needed than a stout stick. "With this," says Azara, "I have killed many by dealing them blows on the head, and with the same security as if I had struck the trunk of a tree. With a mouth so small, and formed as already described, the Ant-Bear cannot bite; and, if it could, it would be useless, for it has no teeth."

"Like a lawyer," says Professor Owen, "the tongue is the chief organ by which this animal obtains its livelihood in its natural habitat. The warmer latitudes of South America, to which part of the world the Ant-Bear is peculiar, abound in forests and luxuriant vegetation; the insects of the ant and termite tribes that subsist on wood, recent or decaying, equally abound. With one link in the chain of organic independencies is interlocked another; and as the surplus vegetation sustains the surplus insect population, so a peculiar form of mammalian life finds the requisite conditions of existence in the task of restraining the undue multiplication of the wood-consuming insects."

The number of male Ant-eaters is supposed to be considerably smaller than that of the females, which circumstance favours the inference that the extinction of the species, like those of the *edentata* in general, is determined upon.*

Large as the Ant-Bear is in comparison with the animals on which it naturally feeds, there appear to have been still larger Ant-Bears in the old times of South America. Fossil remains of nearly allied

* Proceedings of the Zoological Society.

quadrupeds have been detected in both the freshwater deposits and bone-caves of the post-pliocene period in Buenos Ayres and Brazil.

In examining the fossil remains has been found evidence that the nervous matter destined to put in action the muscular part of the tongue was equal to half of that nervous matter which influences the whole muscular system of a man. No other known living animal offers any approximation to the peculiar proportions of the lingual nerves of the fossil animal in question except the Great Ant-eater; but the size of the animal indicated by the fossil was three times that of our Ant-eater. For this strange monster, thus partially restored from the ruins of a former world, Professor Owen proposes the name of *Glossotherium*, which signifies tongue-beast.

Evidence of such a creature has been given by Dr. Lund, the Danish naturalist, resident in Brazil: among the fossil remains here (limestone caves of the province Minas) he discovered traces of the Great Ant-eater, which, however, are too imperfect to enable us to determine more accurately its relation to existing species. The fragments indicate an animal the size of an ox! Were the insect prey of these antediluvian Ant-eaters correspondingly gigantic?

Two circumstances very remarkable were observed in the Zoological Society's Great Ant-eater: the hinge-like manner in which the animal worked its tail when it had laid itself down, throwing it over the whole of its body and enveloping itself completely; and the peculiar vibratory motion of the long vermiform tongue when protruded from the mouth in search of food. The tongue is not shot forth and retracted, like that of the chameleon, but protruded gradually, *vibrating* all the time, and in the same condition withdrawn into the mouth.

Another species of Ant-eater is the *Tamandua*, much inferior in size to the Great Ant-Bear, being scarcely so large as a good sized cat, whilst the other exceeds the largest greyhound in length. The Tamandua inhabits the thick primæval forests of tropical America, and is never found on the ground, but exclusively in trees, where it lives upon termites, honey, and, according to Azara, even bees, which in those countries form their hives among the loftiest branches of the forest; and having no sting, they are more readily despoiled of their honey than their congeners of our own climate. When about to sleep it hides its muzzle in the fur of its breast, falls on its belly, letting its fore-feet hang down on each side, and wrapping the whole tightly round with its tail. The female, as in the Great Ant-eater, has but two pectoral mammæ, and produces but a single cub at a birth, which she carries about with her on her shoulders for the first three or four months. *Tamandua* is the Portuguese name; the French and English call it *fourmiller* and Little Ant-Bear.

The latter are the names of a still smaller species,

which does not exceed the size of the European squirrel. Its native country is Guayana and Brazil. It is called in Surinam *kissing-hand*, as the inhabitants pretend it will never eat, at least when caught, but that it only licks its paws in the same manner as the bear; that all trials to make it eat have proved in vain, and that it soon dies in confinement. Von Sack, in a voyage to Surinam, had two of these Ant-eaters which would not eat eggs, honey, meat, or ants; but when a wasps'-nest was brought they pulled out the nymphæ and ate them eagerly, sitting in the posture of a squirrel. Von Sack showed this phenomenon to many of the inhabitants of Surinam, who all assured him that it was the first time they had ever known that species of animal to take any nourishment.

Von Sack describes his Ant-eaters as often sleeping all the day long curled together, and fastened by their prehensile tails to one of the perches of the cage. When touched they raised themselves on their hind-legs, and struck with their fore-paws at the object which disturbed them, like the hammer of a clock striking a bell, with both paws at the same time, and with a great deal of force. They never attempted to run away, but were always ready for defence when attacked.

The discovery of the true nature of the food of this species is particularly desirable, and may enable us to have the animal brought alive to this country, a thing which we believe has not been attempted; and which, if attempted, has certainly never succceded. To procure or carry ants during a long sea-voyage is impracticable, but the larvæ of wasps can be obtained in any quantity, and will keep for months; so that the most serious difficulty to the introduction of the little Ant-eater being thus removed, it would only require to be protected from the effects of a colder climate, which may be as easily done in its case as in that of other South America mammalia.

The Porcupine Ant-eater of New Holland, now very uncommon in New South Wales, is regarded, of its size, the strongest quadruped in existence. It burrows readily. Its mode of eating is very curious, the tongue being used sometimes in the manner of that of the chameleon, and at other times in that in which a mower uses his scythe, the tongue being curved laterally, and the food, as it were, swept into the mouth.

The original Great Ant-Bear, received at the Gardens of the Zoological Society on the 29th of September, 1853, died on the 6th of July, 1854. There are now two of these animals living in the Gardens, one of which is a remarkably fine specimen.

CURIOSITIES OF BATS.

HESE harmless and interesting little animals have not only furnished objects of superstitious dread to the ignorant, but have proved to the poet and the painter a fertile source of images of gloom and terror. The strange combination of character of beast and bird, which they were believed to possess, is supposed to have given to Virgil the idea of the Harpies.

Aristotle says but little about the Bat; and Pliny is considered to have placed it among the Birds, none of which, he observes, with the exception of the Bat, have teeth. Again, he notices it as the only winged animal that suckles its young, and remarks on its embracing its two little ones, and flying about with them. In this arrangement he was followed by the older of the more modern naturalists. Belon, doubtingly, places it at the end of the Night-birds; and the Bat, *Attaleph* (bird of darkness), was one of the unclean animals of the Hebrews; and in Deuteronomy xxv. 18, it is placed among the forbidden birds.

Even up to a late period Bats were considered as forming a link between quadrupeds and birds. The common language of our own ancestors, however, indicates a much nearer approach to the truth in the notions entertained by the people than can be found in the lucubrations of the learned. The words *reremouse* and *flittermouse*, the old English names for the Bat—the former derived from the Anglo-Saxon "aræan," to raise, or rear up, and mus; the latter from the Belgic, signifying "flying or flittering mouse,"—show that in their minds these animals were always associated with the idea of quadrupeds. The first of these terms is still used in English heraldry; though it may have ceased to belong to the language of the country. "The word *flittermouse*," says Mr. Bell, "sometimes corrupted into *flintymouse*, is the common term for the Bat in some parts of the kingdom, particularly in that part of the county of Kent in which the language, as well as the aspect and names of the inhabitants, retain more of the Saxon character than will be found, perhaps, in any other part of England.

Ben Jonson has—

"Once a Bat, and ever a Bat! a rere-mouse,
And bird o' twilight, he has broken thrice.
 . . .
Come, I will see the flicker-mouse, my fly."
Play.—New Inn.

The same author uses flitter-mouse also:—

"And giddy flitter-mice, with leather wings."
Sad Shepherd.

Calmet describes the Bat as an animal having the body of a mouse and the wings of a bird; but he erroneously adds, "It never grows tame."

Some persons are surprised at Bats being classed by naturalists, not with birds, but quadrupeds. They have, in fact, no other claim to be considered as birds than that of their being able to suspend and move themselves in the air, like some species of fish, but to a greater degree. They suckle their young, are covered with hair, and have no wings, but arms and lengthened fingers or toes furnished with a membrane, by which they are enabled to fly.

Sir Charles Bell, in his valuable treatise on the "Hand," considers the skeleton of the Bat as one of the best examples of the moulding of the bones of the extremity to correspond with the condition of the animal. Contemplating this extraordinary application of the bones of the extremity, and comparing them with those of the wing of a bird, we might say that this is an awkward attempt—"a failure." But, before giving expression to such an opinion, we must understand the objects required in this construction. It is not a wing intended merely for flight, but one which, while it raises the animal, is capable of receiving a new sensation, or sensations, in that exquisite degree, so as almost to constitute a new sense. On the fine web of the Bat's wing nerves are distributed, which enable it to avoid objects in its flight during the night, when both eyes and ears fail. Could the wing of a bird, covered with feathers, do this? Here, then, we have another example of the necessity of taking every circumstance into consideration before we presume to criticise the ways of nature. It is a lesson of humility. In this animal the bones are light and delicate; and whilst they are all marvellously extended, the phalanges of the fingers are elongated so as hardly to be recognised, obviously for the purpose of sustaining a membranous web, and to form a wing.

In 1839 there was received at the Surrey Zoological Gardens, from Sumatra, a specimen of the Vampire Bat. This was a young male; the body was black, and the membranous wing, in appearance, resembled fine black kid. He was rarely seen at the bottom of his cage, but suspended himself from the roof or bars of the cage, head downwards, his wings wrapped round his body; when spread, these wings extended nearly two feet. Although this specimen was the Vampire Bat to which so many bloodthirsty feats have been attributed, his appearance was by no means ferocious; he was active, yet docile, and the only peculiarity to favour belief in his blood-sucking propensity was his long pointed tongue. The species has popularly been accused of destroying, not only the large mammiferous animals, but also men, when asleep, by sucking their blood. "The truth," says Cuvier, in his "Regne Animal," "appears to be, that the Vampire inflicts only small wounds, which may, probably, become inflammatory and gangrenous from the influence of climate." In this habit, however, may have originated the celebrated Vampire superstition. Lord Byron, in his beautiful poem of "The Giaour," thus symbolises the tortures that await the "false infidel:"—

> "First, on earth as Vampire sent,
> My corse shall from its tomb be rent;
> Then ghastly haunt thy native place,
> And suck the blood of all thy race;
> There, from thy daughter, sister, wife,
> At midnight drain the stream of life;
> Yet loathe the banquet which perforce
> Must feed thy livid living corse.
> Thy victims, ere they yet expire,
> Shall know the demon for their sire,
> As cursing thee, thou cursing them,
> Thy flowers are withered on the stem.
> But one that for thy crime must fall,
> The youngest, most beloved of all,
> Shall bless thee with *a father's* name—
> That word shall wrap thy heart in flame!
> Yet must thou end thy task, and mark
> Her cheek's last tinge, her eye's last spark,
> And the last glassy glance must view
> Which freezes o'er its lifeless blue;
> Then with unhallowed hand shall tear
> The tresses of her yellow hair,
> Of which in life a lock, when shorn,
> Affection's fondest pledge was worn,
> But now is borne away by thee,
> Memorial of thine agony!
> Wet with thine one best blood shall drip
> Thy gnashing tooth and haggard lip;
> Then stalking to thy sullen grave,
> Go, and with Gouls and Afrits rave;
> Till there in horror shrink away
> From spectre more accursed than they!"

In a note, the noble poet tells us:—"The Vampire superstition is still general in the Levant." Honest

Tournefort tells a long story, which Mr. Southey, in the notes on "Thalaba," quotes, about these Vardoulacha, as he calls them. "I recollect a whole family being terrified by the screams of a child, which they imagined must proceed from such a visitation. The Greeks never mention the word without horror."

Bishop Heber describes the Vampire Bat of India as a very harmless creature, entirely different from the formidable idea entertained of it in England. "It only eats fruit and vegetables; indeed, its teeth are not indicative of carnivorous habits; and from blood it turns away when offered to it. During the day-time it is, of course, inert; but at night it is lively, affectionate, and playful, knows its keeper, but has no objection to the approach and touch of others."

Mr. Westerton, the traveller, when speaking, in his "Wanderings," of the Vampire of South America, says:—"There are two species in Demerara, both of which suck living animals; one is rather larger than the common Bats, the other measures above two feet from wing to wing, extended. So gently does this nocturnal surgeon draw the blood, that instead of being roused, the patient is lulled into a profound sleep." The large Vampire sucks men, commonly attacking the toes; the smaller seems to confine itself chiefly to birds.

Captain Stedman, who states that he was bitten by a Bat, thus describes the operation:—"Knowing by instinct that the person they intend to attack is in a sound slumber, they generally alight near the feet, where, while the creature continues fanning with its enormous wings, which keeps one cool, he bites a piece out of the tip of the great toe, so very small indeed that the head of a pin would scarcely be received into the wound, which is, consequently, not painful; yet through this orifice he continues to suck the blood until he is obliged to disgorge. He then begins again, and thus continues sucking and disgorging until he is scarcely able to fly; and the sufferer has been often known to sleep from time into eternity. Having applied tobacco-ashes as the best remedy, and washed the gore from myself and my hammock, I observed several small heaps of congealed blood all round the place where I had lain upon the ground, on examining which the surgeon judged that I had lost at least twelve or fourteen ounces during the night."

Lesson, in 1827, says:—" The single American species of Bat is celebrated by the fables with which they have accompanied its history. That Bats suck the blood of animals as well as the juices of succulent fruits zoologists are agreed. The rough tongue of one genus was, I suppose, to be employed for abrading the skin, to enable the animal to suck the part abraded; but zoologists are now agreed that the supposition is groundless. It is more than probable that the celebrated Vampire superstition and the blood-sucking qualities attributed to the Bat have some connection with each other.

Bat-fowling is mentioned by Shakspeare. This is

the mode of taking Bats in the night-time, while they are at roost, upon perches, trees, or hedges. They light torches or straw, and then beat the bushes, upon which the Bats, flying to the flames, are caught, either with nets or otherwise.

Bat-fowling, or Bat-folding, is effected by the use of a net, called a trammel-net, and is practised at night. The net should be made of the strongest and finest twine, and extended between two poles about ten feet high, tapering to a point at the top, and meeting at the top of the net. The larger ends are to be held by the persons who take the management of the net, and who, by stretching out the arms, keep the net extended to the utmost, opposite the hedge in which the Bats or birds are supposed to be. Another of the party carries a lantern upon a pole at a short distance behind the centre of the net. One or two others place themselves on the opposite side of the hedge, and by beating it with sticks disturb the Bats or birds, which, being alarmed, fly towards the light, but are interrupted in their flight by the net which is immediately *folded* upon them, often fifteen or twenty in number. This sport cannot be followed with much success except when the night is very dark, or until very late in the autumn, when the trees, having lost their leaves, the Bats or birds are driven for shelter to the hollies, yews, hay-ricks, &c.

We remember reading, in the "Philosophical Magazine," in 1836, a curious account of the habits of a long-eared Bat, a living specimen of which was given to the children of Mr. De Carle Sowerby, the naturalist. "We constructed," says Mr. Sowerby, "a cage for him, by covering a box with gauze, and making a round hole in the side, fitted with a phial cork. When he was awake, we fed him with flies, introduced through this hole, and thus kept him for several weeks. The animal soon became familiar, and immediately a fly was presented alive at the hole, he would run or fly from any part of the cage, and seize it in our fingers; but a dead or quiet fly he would never touch. At other times, dozens of flies and grasshoppers were left in his cage, and, waking him by their noise, he dexterously caught them as they hopped or flew about, but uniformly disregarded them while they were at rest. The cockroach, hard beetles, and caterpillars he refused.

"As we became still more familiar, our new friend was invited to join in our evening amusements, to which he contributed his full share by flitting round the room, at times settling upon our persons, and permitting us to handle and caress him. He announced his being awake by a shrill chirp, which was more acute than that of the cricket. Now was the proper time for feeding him. I before stated that he only took his food alive. It was observed that not only was motion necessary, but that generally some noise on the part of the fly was required to induce him to accept it; and this fact was soon discovered by the children, who were entertained by his taking flies from their fingers as he flew by them, before he was bold enough to settle upon their hands to devour his victims. They quickly improved upon this discovery, and, by imitating the booming of a bee, induced the Bat, directed by the sound, to settle upon their faces, wrapping his wings round their lips, and searching for the expected fly. We observed that, if he took a fly while on the wing, he frequently settled to masticate it; and, when he had been flying about a long time, he would rest upon a curtain, pricking his ears, and turning his head in all directions, when, if a fly were made to buzz, or the sound imitated, he would proceed directly to the spot, even on the opposite side of the room, guided, it would appear, entirely by the ear. Sometimes he took his victim in his mouth, even though it was not flying; at other times he inclosed it in his wings, with which he formed a kind of bag-net. This was his general plan when in his cage, or when the fly was held in our fingers, or between our lips."

From these observations Mr. Sowerby concludes that many of the movements of the Bat upon the wing are directed by his exquisite sense of hearing. May not the sensibility of this organ be naturally greater in these animals, whose organs of vision are too susceptible to bear daylight, when those organs, from their nature, would necessarily be of most service?—such as the cat, who hunts by the ear, and the mole, who, feeding in the dark recesses of his subterranean abode, is very sensible of the approach of danger, and expert in avoiding it. In the latter case, large external ears are not required, because sound is well conveyed by solids, and along narrow cavities. In the cases of many Bats, and of owls, the external ears are remarkably developed. Cats combine a quickness of sight with acute hearing. They hunt by the ear, but they follow their prey by the eye. Some Bats are said to feed upon fruits: have they the same delicacy of hearing, feeling, &c., as others?

Mr. Sowerby has further described the singular mode adopted by the long-eared Bat in capturing his prey. The flying apparatus is extended from the hind legs to the tail, forming a large bag or net, not unlike two segments of an umbrella, the legs and tail being the ribs. The Bat, having caught the fly, instead of eating it at once, generally covers it with his body, and, by the aid of his arms, &c., forces it into his bag. He then puts his head down under his body, withdraws the fly from his bag, and leisurely devours it. Mr. Sowerby once saw an unwary bluebottle walk beneath the body of the apparently sleeping Bat into the sensitive bag, in which it was immediately imprisoned. White, of Selborne, speaking of a tame Bat, alludes to the above described action, which he compares to that of a beast of prey, but says nothing respecting the bag. Bell, in his "British Quadrupeds," says that the interfemoral membrane of Bats "is probably intended to act as a sort of rudder, in rapidly changing the course of the animal in the pursuit of its insect food. In a large group of foreign Bats, which feed on fruit or other vegetable substances, as well as in some of carnivorous habits, but whose prey is of a less active character, this part is either wholly wanting or much circumscribed in extent and power." May it not be, asks Mr. Sowerby, that they do not require an entomological bag-net?

The wing of the Bat is commonly spoken of as of leather; that it is an insensible piece of stuff—the leather of a glove or of a lady's shoe; but nothing can be further from the truth. If one were to select an organ of the most exquisite delicacy and sensibility, it would be the Bat's wing. It is anything but leather, and is, perhaps, the most acute organ of touch that can be found.

Bats are supposed to perceive external objects without coming actually in contact with them, because in their rapid and irregular flight, amidst various surrounding bodies, they never fly against them; yet, to some naturalists, it does not appear that the senses of hearing, seeing, or smelling serve them on these occasions, for they avoid any obstacles with equal certainty when the eye, ear, and nose are closed: hence has been ascribed a *sixth sense* to these animals. The nerves of the wing are large and numerous, and distributed in a minute network between the integuments. The impulse of the air against this part may possibly be so modified by the objects near which the animal passes as to indicate their situation and nature. The Bat tribe fly by means of the fingers of the fore feet, the thumb excepted, being, in these animals, longer than the whole body; and between them is stretched a thin membrane, or web, for flying. It is probable that, in the action of flight, the air, when struck by this wing, or very sensitive hand, impresses a sensation of heat, cold, mobility, and resistance on that organ, which indicates to the animal the existence or absence of obstacles which would interrupt its progress. In this manner blind men discover by their hands, and even by the skin of their faces, the proximity of a wall, door of a house, or side of a street, even without the assistance of touch, and merely by the sensation which the difference in the resistance of the air occasions. Hence they are as little capable of walking on the ground as apes with their hands, or sloths with their hooked claws, which are calculated for climbing.

In a certain kind of Bat, the *Nycteris*, there exists a power of inflation to such a degree that, when inflated, the animal looks, according to Geoffroy St. Hilaire, like a *little balloon* fitted with wings, a head, and feet. It is filled with air through the cheek-pouches, which are perforated at the bottom, so as to communicate with the spaces of the skin to be filled. When the Bat wishes to inflate, it draws in its breath, closes its nostrils, and transmits the air through the perforations of the cheek-pouches to the spaces; and the air is prevented from returning by the action of a muscle which closes those openings, and by valves of considerable size on the neck and back.

There was formerly a vulgar opinion that Bats, when down on a flat surface, could not get on the wing again, by rising with great ease from the floor; but White saw a Bat run, with more dispatch than he was aware of, though in a most ridiculous and grotesque manner. The adroitness with which this Bat sheared off the wings of flies, which were always rejected, was very amusing. He did not refuse raw flesh when offered; so that the notion that Bats go down chimneys, and gnaw men's bacon, seems no improbable story.

Mr. George Daniell describes a female Bat, who took her food with an action similar to that of a dog. The animal took considerable pains in cleaning herself, parting the hair on either side, from head to tail, and forming a straight line along the middle of the back. The membrane of the wings was cleaned by forcing the nose through the folds, and thereby expanding them. This Bat fed freely, and at some times voraciously, the quantity exceeding half an ounce, although the weight of the animal itself was not more than ten drams.

The *Kalong* Bat of the Javanese is extremely abundant in the lower parts of Java, and uniformly lives in society. The more elevated districts are not visited by it. "Numerous individuals," says Dr. Hornfield, "select a large tree, and, suspending themselves with the claws of their posterior extremities to the naked branches, often in companies of several hundreds, afford to a stranger a very singular spectacle. A species of ficus (fig-tree), resembling the *ficus religiosa* of India, affords them a very favourite retreat, and the extended branches of one of these are sometimes covered by them. They pass the greater portion of the day in sleep, hanging motionless, ranged in succession, with the head downwards, the membrane contracted about the body, and often in close contact. They have little resemblance to living beings; and, by a person not accustomed to their economy, are readily mistaken for a part of the tree, or for a fruit of uncommon size suspended from its branches.

In general, these societies are silent during the day; but if they are disturbed, or a contention arises among them, they emit sharp, piercing shrieks; and their awkward attempts to extricate themselves, when oppressed by the light of the sun, exhibit a ludicrous spectacle. Soon after sunset they gradually quit their hold, and pursue their nocturnal flight in quest of food. They direct their course by an unerring instinct to the forests, villages, and plantations, attacking and devouring every kind of fruit, from the abundant and useful cocoa-nut, which surrounds the dwellings of the meanest peasantry, to the rare and most delicate productions which are cultivated by princes and chiefs of distinction. Various methods are employed to protect the orchards and gardens. Delicate fruits are secured by a loose net or basket, skilfully constructed of split bamboo, without which precaution little valuable fruit would escape the ravages of the *Kalong*. There are few situations in the lower part of Java in which this night wanderer is not constantly observed. As soon as the light of the sun has retired, one animal is seen to follow the other at a small but irregular distance, and this accession continues uninterrupted till dark:—

> "The night came on apace,
> And falling dews bewet around the place;
> The bat takes airy rounds, on leathern wings,
> And the hoarse owl his woful dirges sings."
>
> Gay's "*Pastoral* III."

Bats of the ordinary size are very numerous in Jamaica. They are found in mills and old houses. They do great mischief in gardens, where they eat the green peas, opening the pod over each pea, and removing it very dexterously.

Gilbert White, of Selborne, first noticed a large species of Bat, which he named *altivolans*, from its manner of feeding high in the air. In the extent of its wings it measured $14\frac{1}{2}$ inches; and it weighed, when entirely full, one ounce and one drachm. It is found in numbers together, so many as 185 having been taken in one night from the eaves of Queens' College, Cambridge. In the Northern Zoological Gallery of the British Museum are representatives of the several species of Bats, all bearing a family resemblance to each other. In England alone there are eighteen known species. Here is the curious leaf-nosed Bat, from Brazil, supposed to excel in the sense of smell; also, the Vampire, or large blood-sucking Bat, from the same country; and the different kinds of fruit-eating Bats, found in America and Australia, and sometimes called flying foxes, on account of their great size. The

Bats of temperate climates remain torpid during the winter. Gay has these lines :—

> " Where swallows in the winter season keep ;
> And here the drowsy bat and dormouse sleep."

Young Bats have been taken, when hovering near the ground, by throwing handfuls of sand, but they rarely live in confinement : they often die within a week after their capture. A Bat, taken in Elgin, gave birth to a young one, which was for two days suckled by its parent. Before she reached the age of three days the young bat died, and the parent only survived another day to mourn her loss. Sometimes females, when taken, have young ones clinging to their breast, in the act of sucking ; and the female can fly with ease, though two little ones are attached to her, which weigh nearly as much as the parent.

To return to an exaggeration of a famous old traveller. In " Purchas his Pilgrimage," the materials for which he borrowed from above thirteen hundred authors, when speaking of the island of Madura, in the South of India, he says :—" In these partes are Battes as big as Hennes, which the people roast and eat."

THE HEDGEHOG.

OF this animal some strange things are recorded. It is placed by Cuvier at the head of the insect-devouring Mammifera. It is found in Europe, Africa, and India. Its body is covered with strong and sharp prickles, and by the help of a muscle it can contract itself into a ball, and so withdraw its whole underpart, head, belly, and legs, within this thicket of prickles:

> "Like Hedgehogs, which
> Lie tumbling in my barefoot way, and mount
> Their pricks at my foot-fall."—Shakspeare's "*Tempest.*"

Sir Thomas Browne, in his "Vulgar Errors," has this odd conceit:—"Few have belief to swallow, or hope enough to experience, the collyrium of Albertus; that is, to make one see in the dark: yet thus much, according to his receipts, will the right eye of an Hedgehog, boiled in oil, and preserved in a brazen vessel, effect."

Hedgehog was an old term of reproach; but we have heard a well-set argument compared to a hedgehog—all points.

The food of the Hedgehog, which is a nocturnal animal, consists principally of insects, worms, slugs, and snails. That it will eat vegetables is shown by White of Selborne, who relates how it eats the root of the plantain by boring beneath it, leaving the tuft of leaves untouched.

The Hedgehog is reputed to supply itself with a winter covering of leaves. So far as we are aware, it has not been observed in the act of forming the covering of leaves, though it is supposed to roll itself about till its spines take up a sufficient number, in the same way as it is popularly believed (without proof) to do with apples. Blumenbach states that he was assured, " by three credible witnesses," that Hedgehogs so gather fruit; but Buffon, who kept several Hedgehogs for observation, declares they never practise any such habit.

The voracity of the Hedgehog is very great. A female, with a young one, was placed in a kitchen, having the run of the beetles at night, besides having always bread and milk within their reach. One day, however, the servants heard a mysterious crunching sound in the kitchen, and found, on examination, that nothing was left of the young Hedgehog but the skin and prickles—the mother had devoured her little pig! A Hedgehog has also been known to eat a couple of rabbits which had been confined with it, and killing others; it has likewise been known to kill hares.

A Hedgehog was placed in one hamper, a wood-pigeon in another, and two starlings in a third;

the lid of each hamper was tied down with string, and the hampers were placed in a garden-house, which was fastened in the evening. Next morning the strings to the hampers were found severed, the starlings and wood-pigeon dead and eaten, feathers alone remaining in their hampers, and the Hedgehog alive in the wood-pigeon's hamper. As no other animal could have got into the garden-house it was concluded that the Hedgehog had killed and eaten the birds.

In the "Zoological Journal," vol. ii., is an account by Mr. Broderip of an experiment made by Professor Buckland proving that in captivity at least the Hedgehog will devour snakes; but there is no good reason for supposing that it will not do the same in a state of nature, for frogs, toads, and other reptiles, and mice, have been recorded as its prey. From its fondness for insects it is often placed in the London kitchens to keep down the swarm of cockroaches with which they are infested; and there are generally Hedgehogs on sale at Covent Garden Market for this purpose.

The idle story that the persecuted Hedgehog sucks cows has been thus quaintly refuted :—" In the case of an animal giving suck, the teat is embraced round by the mouth of the young one, so that no air can pass between; a vacuum is made, or the air is exhausted from its throat, by a power in the lungs; nevertheless the pressure of the air remains still upon the outside of the dug of the mother, and by these two causes together the milk is forced in the mouth of the young one. But a Hedgehog has no such mouth as to be able to contain the teat of a cow; therefore any vacuum which is caused in its own throat cannot be communicated to the milk in the dug. And if he is able to procure no other food but what he can get by sucking cows in the night, there is likely to be a vacuum in his stomach too." (*New Catalogue of Vulgar Errors.* By Stephen Fovargue, A.M., 1786.) Yet, according to Sir William Jardine, the Hedgehog is very fond of eggs; and is consequently very mischievous in the game-preserve and hen-house.

One of the most interesting facts in the natural history of the Hedgehog is that announced in 1831 by M. Lenz, and subsequently confirmed by Professor Buckland: this is, that the most violent poisons have no effect upon it; a fact which renders it of peculiar value in forests, where it appears to destroy a great number of noxious reptiles. M. Lenz says that he had in his house a female Hedgehog, which he kept in a large box, and which soon became very mild and familiar. He often put into the box some adders, which it attacked with avidity, seizing them indifferently by the head, the body, and the tail, and not appearing alarmed or embarrassed when they coiled themselves around its body. On one occasion M. Lenz witnessed a fight between a Hedgehog and a viper. When the Hedgehog came near and smelled the snake, for with these animals the sense of sight is very obtuse, she seized it by the head, and held it fast between her teeth, but without appearing to do it much harm; for having disengaged its head, it assumed a furious and menacing attitude, and, hissing vehemently, inflicted severe bites on the Hedgehog. The animal did not, however, recoil from the bites of the viper, or indeed seem to care much about them. At last, when the reptile was fatigued by its efforts, she again seized it by the head, which she ground beneath her teeth, compressing the fangs and glands of poison, and then devouring every part of the body. M. Lenz says that battles of this sort often occurred in the presence of many persons, and sometimes the Hedgehog received eight or ten wounds on the ears, the snout, and even on the tongue, without seeming to experience any of the ordinary symptoms produced by the venom of the viper. Neither herself nor the young which she was then suckling seemed to suffer from it. This observation agrees with that of Pallas, who assures us that the Hedgehog can eat about a hundred Cantharides (Spanish Flies) without experiencing any of the effects which this insect, taken inwardly, produces on men, dogs, and cats. A German physician, who made the Hedgehog a particular object of study, gave it strong doses of prussic acid, of arsenic, of opium, and of corrosive sublimate, none of which did it any harm. The Hedgehog in its natural state only feeds on pears, apples, and other fruits when it can get nothing it likes better.

The Hedgehog hybernates regularly, and early in

the summer brings forth from two to four young ones at a birth, which, at the time of their production, are blind, and have the spines white, soft, and flexible. The nest wherein they are cradled is said to be very artificially constructed, the roof being rain-proof.

The flesh of the Hedgehog, when it has been well fed, is sweet and well flavoured, and is eaten on the Continent in many places. In Britain a few besides gipsies partake of it. The prickly skin appears to have been used by the Romans for hackling hemp.

Gilbert White notes that when the Hedgehog is very young it can draw its skin down over its face, but is not able to contract itself into a ball, as the creature does, for the sake of defence when full grown. The reason, White supposes, is because the curious muscle that enables the Hedgehog to roll itself up into a ball has not then arrived at its full tone and firmness. Hedgehogs conceal themselves for the winter in their warm *hybernaculum* of leaves and moss; but White could never find that they stored in any winter provision, as some quadrupeds certainly do.

THE HIPPOPOTAMUS IN ENGLAND.

IN the year 1850 there was exhibited in London a living Hippopotamus, for many centuries the only instance of this extraordinary animal being seen in Europe.

There is something irresistibly striking in seeing a living animal, not one of whose species we have before seen, and especially when that animal is a large one, as in the instance before us. We had been wonderstruck at forms of this creature in the old British Museum, where were two finely-preserved specimens. The Rhinoceros alive was; until of late years, very rare in England. In 1834 Mr. Cross paid some 1,500*l*. for a young Indian one-horned Rhinoceros, this being the only one brought to England for twenty years. He proved attractive, but slightly so in comparison with the expectation of a living Hippopotamus, never witnessed before in this country. The circumstances of his acquisition were as follows :—

The Zoological Society of London had long been

anxious to obtain a living Hippopotamus for their menagerie, but without success. An American agent at Alexandria had offered 5,000*l*. for an animal of this species, but in vain; no speculator could be induced to encounter the risk and labour of an expedition to the White Nile for the purpose of securing the animal. The desire of the Zoological Society was communicated to the Viceroy of Egypt, who saw the difficulty. Hasselquist states it to have been impossible to bring the living animal to Cairo; and the French *savans*, attached to the expedition to Egypt, who ascended the Nile above Syene, did not meet with one Hippopotamus. Caillaud, however, asserts that he saw forty Hippopotami in the Upper Nile, though their resort lay fifteen hundred miles or more from Cairo. Here they were often shot with rifle-balls, but to take one alive was another matter. However, by command of the Viceroy, the proper parties were sent in search of the animal.

In August, 1849, the hunters having reached the island of Fobaysch, on the White Nile, about 2,000 miles above Cairo, shot a large female Hippopotamus in full chase up the river. The wounded creature turned aside and made towards some bushes on the island bank, but sank dead in the effort. The hunters, however, kept on towards the bushes, when a young Hippopotamus, supposed to have been recently brought forth, not much bigger than a new-born calf, but stouter and lower, rushed down the bank of the river, was secured by a boatman and lifted into the boat. The captors started with their charge down the Nile. The food of their young animal was their next anxiety; he liked neither fish, flesh, fruit, nor grass. The boat next stopped at a village; their cows were seized and milked, and the young charge lapped up the produce. A good milch cow was taken on board, and with this supply the Hippopotamus reached Cairo. The colour of his skin at this time was a dull reddish brown. He was shown to the Pasha in due form; the present created intense wonder and interest in Cairo; gaping crowds filled its narrow sandy streets, and a whale at London-bridge would scarcely excite half so much curiosity.

It being thought safer for the animal to winter in Cairo than to proceed forthwith on his journey, the Consul had duly prepared to receive the young stranger, for whom he had engaged a sort of nurse, Hamet Safi Cannana. An apartment was allotted to the Hippopotamus in the court-yard of the Consul's house, leading to a warm or tepid bath. His milk-diet, however, became a troublesome affair, for the new comer never drank less than from twenty to thirty quarts daily.

By the next mail after the arrival of the Hippopotamus, the Consul despatched the glad tidings to the Zoological Society. The animal was shipped at Alexandria, in the Ripon steamer. On the main deck was built a house, from which were steps down into an iron tank in the hold, containing 400 gallons of water, as a bath: it was filled with fresh water every other day.

Early in May, the Hippopotamus was conveyed in the canal-boat, with Hamet Safi Cannana, to Alexandria, where the debarkation was witnessed by 10,000 spectators. The animal bore the voyage well. He lived exclusively on milk, of which he consumed daily about forty pints, yielded by the cows taken on board. He was very tame, and, like a faithful dog, followed his Arab attendant Hamet, who was seldom away more than five minutes without being summoned to return by a loud grunt. Hamet slept in a berth with the Hippopotamus. On May 25 they were landed at Southampton, and sent by railway to London. On arriving at the Zoological Society's Gardens, Hamet walked first out of the transport van, with a bag of dates over his shoulder, and the Hippopotamus trotted after him. Next morning he greatly enjoyed the bath which had been prepared for him. Although scarcely twelve months old, his massive proportions indicated the enormous power to be developed in his maturer growth; while the grotesque expression of his physiognomy far exceeded all that could be imagined from the stuffed specimens in museums, and the figures which had hitherto been published from the reminiscences of travellers.

Among the earliest visitors was Professor Owen, who first saw the Hippopotamus lying on its side in the straw, with its head resting against the chair in which sat the swarthy attendant. It now and then grunted softly, and, lazily opening its thick, smooth eyelids, leered at its keeper with a singular protruding movement of the eyeball from the prominent socket, showing an unusual proportion of the white. The retraction of the eyeball was accompanied by a simultaneous rolling obliquely downwards, or inwards, or forwards. The young animal, then ten months' old, was seven feet long, and six and a-half in girth at the middle of the barrel-shaped trunk, supported, clear of the ground, on very short and thick legs, each terminated by four spreading hoofs, the two middle ones being the largest, and answering to those in the hog. The naked hide, covering the broad back and sides, was of a dark, india-rubber colour, with numerous fine wrinkles crossing each other, but disposed almost transversely. The beast had just left its bath, when a glistening secretion gave the hide, in the sunshine, a very peculiar aspect. When the animal was younger, the secretion had a reddish colour, and the whole surface of the hide became painted over with it every time he quitted his bath.

The ears, which were very short, conical, and fringed with hairs, it moved about with much vivacity. The skin around them was of a light reddish-brown colour, and almost flesh-coloured round the eyelids, which defended the prominent eyes, which had a few short hairs on the margin of the upper lid. The colour of the iris was of a dark brown. The nostrils, situated on prominences, which the animal had the power of raising on the upper part of the broad and massive muzzle, were short oblique slits, guarded by two valves, which were opened and closed spontaneously, like the eyelids. The movements of these apertures were most conspicuous when the beast was in the bath.

The wide mouth was chiefly remarkable for the upward curve of its angles towards the eyes, giving a quaintly comic expression to the massive countenance. The short and small milk-tusks projected a little, and the minute incisors appeared to be sunk in pits of the thick gums; but the animal would not permit any close examination of the teeth, withdrawing his head from the attempt, and then threatening to bite. The muzzle was beset with short bristles, split into tufts or pencils of hairs; and fine and short hairs were scattered all over the back and sides. The tail was not long, rather flattened and tapering to an obtuse point.

We may here observe that, at certain moments, the whole aspect of the head suggested to one the idea of what may have been the semblance of some of the gigantic extinct Batrachians (as sirens), the relics of a former world, whose fossil bones in the galleries of Palæontology in the British Museum excite our special wonder.

After lying about an hour, now and then raising its head, and swivelling its eyeballs towards the keeper, or playfully opening its huge mouth, and threatening to bite the leg of the chair on which the keeper sat, the Hippopotamus rose, and walked very

H

slowly about its room, and then uttered a loud and short harsh snort four or five times in quick succession, reminding one of the snort of a horse, and ending with an explosive sound, like a bark. The keeper understood the language—the animal desired to return to its bath.

The Hippopotamus carried its head rather depressed, reminding one of a large prize hog, but with a breadth of muzzle and other features peculiarly its own. The keeper opened the door leading into a paddock, and walked thence to the bath, the Hippopotamus following, like a dog, close to his heels. On arriving at the bath-room, the animal descended with some deliberation the flight of low steps leading into the water, stooped and drank a little, dipped his head under, and then plunged forwards. The creature seemed inspired with new life and activity. Sinking to the bottom of the bath, and moving about submerged for a while, it suddenly rose with a bound almost bodily out of the water. Splashing back, it commenced swimming and plunging about, rolling from side to side, taking in mouthfuls of water and spirting them out again, raising every now and then its huge and grotesque head, and biting the woodwork of the margin of the bath. The broad rounded back of the animal being now chiefly in view, it seemed a much larger object than when out of the water.

After half an hour spent in this amusement, the Hippopotamus quitted the water at the call of its keeper, and followed him back to the sleeping-room, which was well bedded with straw, and where a stuffed sack was provided for its pillow, of which the animal, having a very short neck, thicker than the head, availed itself when it slept. When awake, it was very impatient of any absence of its favourite attendant. It would rise on its hind legs, and threaten to break down the wooden fence, by butting and pushing against it in a way very significant of its great muscular force. The animal appeared to be in perfect health, and breathed, when at rest, slowly and regularly, from three to four times in a minute. Its food was now a kind of porridge, of milk and maize-meat, it being more than half weaned from milk diet. Its appetite had been in no respect diminished by the confinement and inconvenience of the sea voyage, or by change of climate. All observers appear to have agreed that, to see the Hippopotamus rightly, is to see him in the water. There his activity is only surpassed by that of the otter or the seal. Such was one of the opportunities afforded to zoologists for "studying this most remarkable and interesting African mammal, of which no living specimen had been seen in Europe since the period when Hippopotami were last exhibited by the third Gordian in the amphitheatre of imperial Rome."*

It is now time to glance at the general economy of the Hippopotamus, as he is seen in his native rivers and wilds. In early days, as his Roman name imports, it was usual to consider him as a

* Professor Owen.

species of horse, inhabiting rivers and marshy grounds, and, in a more especial manner, the denizen of the Nile. The genus is placed by Linnæus among his *belluæ*, between *equus* and *sus*. The skeleton approaches that of the ox and of the hog, but it presents differences from that of any other animal.

The Hippopotamus is found not only in the Nile, but in the rivers of southern Africa. In the former stream of marvels, Hasselquist relates that "the oftener the River Horse goes on shore, the better hope have the Egyptians of a sufficient swelling or increase of the Nile." Again, they say that the River Horse is an inveterate enemy to the crocodile, and kills it whenever he meets it; adding that he does much damage to the Egyptians in those places he frequents. He goes on shore, and, in a short space of time, destroys an entire field of corn or clover, not leaving the least verdure, for he is very voracious.

Yet neither of these stories is so marvellous as that which a sailor related to Dampier, the old traveller:—"I have seen," says the mariner, "one of these animals open its jaws, and, seizing a boat between its teeth, at one bite sink it to the bottom. I have seen it, on another occasion, place itself under one of our boats, and, rising under it, overset it with six men who were in it, but who, however, happily received no other injury."

Professor Smith and Captain Tuckey, in exploring the Congo River, in South Africa, saw in a beautiful sandy cove, at the opening of a creek, behind a long projecting point, an immense number of Hippopotami; and in the evening a number of alligators were also seen there; an association hardly consistent with the hostility related by Hasselquist.

Captain Tuckey observed Hippopotami with their heads above the water, "snorting in the air." In another part of his narrative he says:—"Many Hippopotami were visible close to our tents at Condo Yanga. No use firing at these animals in the water; the only way is to wait till they come on shore to feed at night."

Le Vaillant had an opportunity of watching the progress of a Hippopotamus under water at Great River, which contained many of these animals. On all sides he could hear them bellow and blow. Anxious to observe them, he mounted on the top of an elevated rock which advanced into the river, and he saw one walking at the bottom of the water. Le Vaillant killed it at the moment when it came to the surface to breathe. It was a very old female, and many people, in their surprise, and to express its size, called it the Grandmother of the River.

The traveller Lander tells us that, on the Niger, Hippopotami are termed water-elephants. One stormy night, as they were sailing up this unexplored current, they fell in with great numbers of Hippopotami, who came plashing, snorting, and plunging all round the canoe. Thinking to frighten them off, the travellers fired a shot or two at them, but the noise only called up from the water and out of the fens about as many more Hippopotami, and they were more closely beset than before. Lander's people, who had never, in all their lives, been exposed to such formidable beasts, trembled with fear, and absolutely wept aloud; whilst peals of thunder rattled over their heads, and the most vivid lightning showed the terrifying scene. Hippopotami frequently upset canoes in the river. When the Landers fired, every one of them came to the surface of the water, and pursued them over to the north bank. A second firing was followed by a loud roaring noise. However, the Hippopotami did the travellers no kind of mischief whatever.

Captain Gordon, when among the Bakalahari, in South Africa, bagged no fewer than fifteen first-rate Hippopotami; the greater number of them being bulls.

In 1828, there was brought to England the head of a Hippopotamus, with all the flesh about it, in high preservation. The animal was harpooned while in combat with a crocodile in a lake in the interior of Africa. The head measured nearly four feet in length, and eight feet in circumference; the jaws opened two feet, and the cutting teeth, of which it had four in each jaw, were above a foot long, and four inches in circumference.

The utility of this vast pachydermatous, or thick-skinned animal, to man is considerable. That he can be destructive has already been shown in his clearance of the cultivated banks of rivers. The enormous ripping, chisel-like teeth of the lower jaw fit him for uprooting. The ancient Egyptians held the animal as an emblem of power, though this may have arisen from his reputed destruction of the crocodile. The flesh is much esteemed for food, both among the natives and colonists of South Africa. The blood of the animal is said to have been used by the old Indian painters in mixing their colours. The skin is extensively employed for making whips.

But there is no part of the Hippopotamus more in request than the great canine teeth, the ivory of which is so highly valued by dentists for making artificial teeth, on account of its keeping its colour better than any other kind. This superiority was not unknown to the ancients Pausanias mentions the statue of Dindymene, whose face was formed of the teeth of Hippopotami, instead of elephants' ivory. The canine teeth are imported in great numbers into England, and sell at a very high price. From the closeness of the ivory, the weight of the teeth, a part only of which is available for the artificial purpose above mentioned, is great in proportion to its bulk; and the article has fetched about thirty shillings per pound.

The ancient history of the Hippopotamus is extremely curious, and we have many representations of him in coins, in sculpture, and in paintings, which prove, beyond question, that the artists, as well as the writers, had a distinct knowledge of what they intended to represent.

The earliest notice which occurs in any author, and which has been considered by many to be a description of the Hippopotamus, is the celebrated account in the fortieth and forty-first chapter of the Book of

Job of Behemoth and Leviathan. Many learned men have contended that "Behemoth" really means "Elephant," and thus the Zurich version of the Bible translates the Hebrew by "Elephas."

In the edition of the English Bible, printed by Robert Barker, in 1615, for King James I., and since considered as the authorised version, the word "Behemoth" is preserved in the text, and the following annotation is added:—" This beast is thought to bee the Elephant, or some other which is unknowen." Bochart, Ludolph, and some others, have contended warmly in favour of the Hippopotamus. Cuvier thinks, that though this animal is probably intended, yet that the description is too vague for any one to hold a certain opinion on the subject. The theory started by Bochart, and in the main supported by Cuvier, is generally supposed the real one. The description in the Book of Job, though doubtless vague, and in the highest degree poetical, has yet sufficient marks to render the identification perfectly easy, while there are certain peculiarities mentioned, which even a poetical imagination could hardly apply to the Elephant. Thus, when it is said of him, " He lieth under the shady trees, in the desert of the reed and fens; . . . the willows of the brook compass him round about," this would seem to be the description of an animal which frequented the water much more than Elephants are accustomed to do. Again, in the fuller description of "Leviathan," in the forty-first chapter, we think it is quite clear that a water animal is intended, though what is there stated might be held to apply to the crocodile as well as the Hippopotamus; both are animals remarkable for extreme toughness of skin, and both are almost equally difficult to kill or to take alive.

Of profane authors, Herodotus is the first who notices this animal, but his account is far from accurate: the size he states as large as the biggest ox. That the animal was sacred, in some parts at least, appears from Herodotus, who says:—"Those which are found in the district of Paprennis are sacred, but in other parts of Egypt they are not considered in the same light." Aristotle makes it no bigger than an ass; Diodorus, an elephant; Pliny ascribes to it the tail and teeth of a boar, adding, that helmets and bucklers are made of the skin. Hippopotami figured in the triumphal processions of the Roman conquerors on their return home. M. Scaurus exhibited five crocodiles and an Hippopotamus; and Augustus one in his triumph over Cleopatra. Antoninus exhibited Hippopotami, with lions and other animals; Commodus no less than five, some of which he slew with his own hand. Heliogabalus, and the third Gordian, also exhibited Hippopotami.

The Hippopotamus of the London Zoological Society was joined by his mate, the more juvenile "Adhela," in 1853. Two Hippopotami have lately been born in Europe; one in the Garden of Plants, at Paris, in 1858; and another in the Zoological Gardens at Amsterdam, in 1866.

With regard to the alleged disappearance of the

Hippopotamus from Lower Egypt, Cuvier remarks, that the French savans attached to the Expedition to Egypt, who ascended the Nile above Syene, did not meet with one.

In some of the rivers of Liberia, and other parts, perhaps, of Western Africa, a second species of Hippopotamus exists, and is proved to be a very distinct animal.

We have yet to glance at the Hippopotami of a former world. Many species are recognised in the fossil remains of Europe and Asia as formerly existing in England and in France. Cuvier detected bones of the Hippopotamus among the fossil wealth of the Great Kirkdale Cavern in Yorkshire, in 1821. They have also been found in France, and especially in the Sewatick Hills in India.

In the Museum of the London Zoological Society are two skulls of Hippopotami—one fossil. This measures two feet three inches, and allowing for skin and lip, two feet six inches. Now, as the head is about one-fifth the length of the body, without the tail, the full-grown animal would be little, if any, short of fifteen feet from nose to tail—a size worthy the description of the Behemoth.

We may here add, that Burckhardt, in his "Travels in Nubia," describes the voice of the Hippopotamus as a hard and heavy sound, like the creaking or groaning of a large wooden door. This noise, he says, is made when the animal raises his huge head out of the water, and when he retires into it again.

LION-TALK.

THE Lion has, within the present century, lost caste, and fallen considerably from his high estate. He has been stripped of much of his conventional reputation by the spirit of inquiry into the validity of olden notions, which characterises the present age; and it appears that much of his celebrity is founded upon popular error. Nor are these results the work of stay-at-home travellers; but they are derived from the observation and experience of those who, amidst scenes of perilous adventure, seek to enlarge and correct our views of the habit and character of the overrated Lion.

Mr. Bennett, in his admirable work, "The Tower Menagerie," has these very sensible remarks:—"In speaking of the Lion we call up to our imaginations the splendid picture of might unmingled with ferocity, of courage undebased by guile, of dignity tempered by grace and ennobled by generosity. Such is the Lion of Buffon; who, in describing this animal, as in too many other instances, has suffered himself to be borne along by the strong tide of popular opinion; but, as the Lion appears in his native regions, according to the authentic accounts of those travellers and naturalists who have had the best means of correctly observing his habits, he is by no means so admirable a creature. Where the timid antelope and ·powerless monkey fall his easy and unresisting prey—or where the elephant and buffalo find their unwieldy bulk and strength no adequate protection against his impetuous agility— he stalks boldly to and fro in fearless majesty. But in the neighbourhood of man—even in that of uncultivated savages—*he skulks in treacherous ambush for his prey.* Of his forbearance and· generosity it can merely be said, that when free, he destroys only what is sufficient to satiate his hunger or revenge; and when in captivity—his wants being provided for, and his feelings not irritated—he suffers smaller animals to live unmolested in his den, or submits to the control of a keeper by whom he is fed. But even this limited degree of docility is liable to fearful interruptions from the calls of hunger, the feelings of revenge—and these he frequently cherishes for a long period—with various other circumstances which render it dangerous to approach him in his most domesticated state, without ascertaining his immediate mood and temper. That an animal which seldom attacks by open force, but silently approaches his victim, and when he imagines his prey to be within his reach, bounds upon it with an overwhelming leap, should ever have been regarded as the type of courage and the emblem of magnanimity, is indeed most astonishing!"

The generosity of disposition so liberally accorded to this powerful beast has been much and eloquently praised; and it seems hard to dissipate the glowing vision which Buffon has raised; but, if there is any dependence to be placed on the observations of those travellers who have had the best opportunities of judging, and have the highest character for veracity, we must be compelled to acknowledge that Buffon's Lion is the Lion of poetry and prejudice, and very unlike the cautious lurking savage that steals on its comparatively weak prey by surprise, overwhelms it at once by the terror, the weight, and the violence of the attack, and is intent only on the gratification of the appetite. "At the time," says Mr. Burchell, "when men first adopted the Lion as the emblem of courage, it would seem that they regarded great size and strength as indicating it; but they were greatly mistaken in the character they had given of the indolent animal." Indeed, Mr. Burchell calls the Lion an "indolent skulking animal." The fact of the Lion sparing the dog that was thrown to him, and making a friend of the little animal that was destined for his prey, has been much dwelt on; but these and other such acts of mercy, as they have been called, may be very easily accounted for. If not pressed by hunger, the Lion will seldom be at the trouble of killing prey; and the desire for a companion has created much stronger friendships between animals in confinement than between a

Lion and a little dog. St. Pierre touchingly describes the Lion of Versailles, who, in 1792, lived most happily with a dog, and on whose death he became disconsolate and miserable; and in confinement the "lordly Lion," as Young calls him, has been known to be deeply afflicted with melancholy at similar losses.

The Lion is easily tamed, and capable of attachment to man. The story of Androdas, frequently called Androcles, is too well known to need more than allusion; but in this and other stories of Lions licking men's hands without injuring them, there must be a stretch of fancy; for the Lion's tongue has sharp thorn-points, inclining backwards, so as not to be able to lick the hand without tearing away the skin, which any one will understand who has *heard* the Lion tear the raw meat away from the bone of his food.

Still, very different accounts are given by travellers of the cruelty or generosity of the Lion's nature; which results, in all probability, from a difference in time or circumstances, or the degree of hunger which the individual experienced when the respective observations were made upon him.

Meanwhile, there are many points in the history of the Lion which are yet but imperfectly understood; the explanations of which, whilst they are interesting, add to our correct knowledge of this still extraordinary animal.

The Lion has been styled "The King of the Forest," which is not very applicable to him, seeing that Mr. Burchell at least never met with but one Lion on the plains; nor did he ever meet with one in any of the forests where he had been. The low cover that creeps along the sides of streams, the patches that mark the springs in the rank grass of the valley, seem to be the shelter which the African Lion, for the most part, seeks. His strength is extraordinary. To carry off a man (and there are dismal accounts of this horrible fact, which there is no reason to doubt) appears a feat of no difficulty to this powerful brute. A Cape Lion, seizing a heifer in his mouth, has carried her off with the same ease as a cat does a rat; and has leaped with her over a broad dyke without the least difficulty. A young Lion, too, has conveyed a horse about a mile from the spot where he had killed it.

There seems to be an idea that the Lion preserves human prey; but, be this as it may, the inhabitants of certain districts have been under the necessity of resorting to a curious expedient to get out of the Lion's reach. Ælian, by the way, records the extinction of a Libyan people by an invasion of Lions. We read of a large tree, in the country of the Mantatees, which has amidst its limbs fourteen conical huts. These are used as dormitories, being beyond the reach of the Lions, which, since the incursions of the Mantatees, when so many thousands of persons were massacred, have become very numerous in the neighbourhood, and destructive to human life. The branches of the above trees are supported by forked sticks or poles, and there are three tiers or platforms on which the huts are constructed. The lowest is nine feet from the ground, and holds ten huts; the second, about eight feet high, has three huts; and the upper story, if it may be so called, contains four. The ascent to these is made by notches cut in the poles; the huts are built with twigs, and thatched with straw, and will contain two persons conveniently. This tree stands at the base of a range of mountains due east of Kurrichaine, in a place called "Ongorutcie Fountain," about 1,000 miles northeast of Cape Town. Kurrichaine is the Staffordshire as well as the Birmingham of that part of South Africa. There are likewise whole villages of huts erected on stakes, about eight feet from the ground; the inhabitants, it is stated, sit under the shade of these platforms during the day, and retire to the elevated huts at night.

Though mortal accidents frequently occur in Lion-hunting, the cool sportsman seldom fails of using his rifle with effect. Lions, when roused, it seems, walk off quietly at first, and if no cover is near, and they are not pursued, they gradually mend their pace to a trot, till they have reached a good distance, and then they bound away. Their demeanour is careless, as if they did not want a fray, but if pressed, are ready to fight it out. If they are pursued closely, they turn and crouch, generally with their faces to the adversary: then the nerves of the sportsman are tried. If he is collected, and master of his craft, the well-directed rifle ends the scene at once; but if, in the flutter of the moment, the vital parts are missed, or the ball passes by, leaving the Lion unhurt, the infuriated beast frequently charges on his enemies, dealing destruction around him. This, however, is not always the case; and a steady, unshrinking deportment has, in some instances, saved the life of the hunter.

There is hardly a book of African travels which does not teem with the dangers and hair-breadth escapes of the Lion-hunters; and hardly one that does not include a fatal issue to some engaged in this hazardous sport. The modes of destruction employed against the powerful beast are very various—from the poisonous arrow of the Bushman to the rifle of the colonist.

The Lion may be safely attacked while sleeping, because of the dullness of his sense of hearing, the difficulty of awakening him, and his want of presence of mind if he be so awakened. Thus the Bushmen of Africa are enabled to keep the country tolerably clear of Lions, without encountering any great danger. The bone of the Lion's fore-leg is of remarkable hardness, from its containing a greater quantity of phosphate of lime than is found in ordinary bones, so that it may resist the powerful contraction of the muscles. The texture of this bone is so compact that the substance will strike fire with steel. He has little sense of taste, his lingual or tongue-nerve not being larger than that of a middle-sized dog.

The true Lions belong to the Old World exclusively, and they were formerly widely and abundantly diffused; but at present they are confined to

Asia and Africa, and they are becoming every day more and more scarce in those quarters of the globe. That Lions were once found in Europe there can be no doubt. Thus it is recorded by Herodotus that the baggage-camels of the army of Xerxes were attacked by Lions in the country of the Reonians and the Crestonæi on their march from Acanthus (near the peninsula of Mount Athos) to Therma, afterwards Thessalonica (now Saloniki); the camels alone, it is stated, were attacked, other beasts remaining untouched, as well as men. Pausanias copies the above story, and states, moreover, that Lions often descended into the plains at the foot of Olympus, which separate Macedonia from Thessaly, and that Polydamas, a celebrated athlete, slew one of the Lions, although he was unarmed.

Nor is Europe the only part of the world from which the form of the Lion has disappeared. Lions are no longer to be found in Egypt, Palestine, or Syria, where they once were evidently far from uncommon. The frequent allusions to the Lion in the Holy Scriptures, and the various Hebrew terms there used to distinguish the different ages and sex of the animal, prove a familiarity with the habits of the race. Even in Asia generally, with the exception of some countries between India and Persia and some districts of Arabia, these magnificent beasts have, as Cuvier observes, become comparatively rare, and this is not to be wondered at. To say nothing of the immense draughts on the race for the Roman arena,—and they were not inconsiderable, for, as Zimmerman has shown, there were 1,000 lions killed at Rome in the space of forty years,—population and civilization have gradually driven them within narrower limits, and their destruction has been rapidly worked in modern times, when firearms have been used against them instead of the bow and the spear. Sylla gave a combat of one hundred Lions at once in his ædileship; but this exhibition is insignificant when compared with those of Pompey and Cæsar, the former of whom exhibited a fight of six hundred, and the latter of four hundred Lions. In Pompey's show three hundred and fifteen of the six hundred were males. The early Emperors consumed great numbers, frequently a hundred at a time, to gratify the people.

The African Lion is annually retiring before the persecution of man farther and farther from the Cape. Mr. Bennett says of the Lion:—"His true country is Africa, in the vast and untrodden wilds of which, from the immense deserts of the north to the trackless forests of the south, he reigns supreme and uncontrolled. In the sandy deserts of Arabia, in some of the wild districts of Persia, and in the vast jungles of Hindostan, he still maintains a precarious footing; but from the classic soil of Greece, as well as from the whole of Asia Minor, both of which were once exposed to his ravages, he has been entirely dislodged and extirpated."

Niebuhr places Lions among the animals of

Arabia; but their proper country is Africa, where their size is the largest, their numbers are greatest, and their rage more tremendous, being inflamed by the influence of a burning sun upon a most arid soil. Dr. Fryer says that those of India are feeble and cowardly. In the interior parts, amidst the scorched and desolate deserts of Zaara or Biledugerid, they reign the masters; they lord it over every beast, and their courage never meets with a check where the climate keeps mankind at a distance. The nearer they approach the habitations of the human race the less their rage, or rather the greater is their timidity: they have often had experienced unequal combats, and finding that there exists a being superior to themselves, commit their ravages with more caution; a cooler climate, again, has the same effect, but in the burning deserts, where rivers and springs are denied, they live in a perpetual fever, a sort of madness fatal to every animal they meet with.

The watchfulness and tenacity of the Lion for human prey are very extraordinary. Mr. Barrow relates that a Lion once pursued a Hottentot from a pool of water, where he was driving his cattle to drink, to an olive-tree, in which the man remained for twenty-four hours, while the Lion laid himself at the foot of the tree. The patience of the beast was at length worn out by his desire to drink, and while he satisfied his thirst the Hottentot fled to his house, about a mile off.

The Lion, however, returned to the tree, and tracked the man within three hundred yards of his dwelling.

Dr. Philip relates a horrible story of a very large Lion recorded at Cape Town in the year 1705. He was known to have seized a sentry at a tent, and was pursued and fired at by many persons without effect. Next morning the Lion walked up a hill *with the man in his mouth,* when about forty shots were fired at him without hitting him; and it was perceived by the blood, and a piece of the clothes of the sentry, that the Lion had taken him away and carried him with him. He was pursued by a band of Hottentots, one of whom he seized with his claws by the mantle, when the man stabbed him with an assagai. Other Hottentots adorned him with their assagais, so that he looked like a porcupine; he roared and leaped furiously, but was at length shot dead. He had a short time before carried off a Hottentot and devoured him.

The Bengal or Asiatic Lion is distinguished from that of Southern Africa principally by the larger size, the more regular and graceful form, the generally darker colour, and the less extensive mane than the African. William Harvey, the graceful artist, drew a portrait of a very fine Bengal Lion, little more than five years old, and then in the Tower collection, and called by the keepers "the Old Lion;" the magnificent development of the mane is very striking in this figure.

 · Maneless Lions have been found on the confines

of Arabia, and were known to Aristotle and Pliny; a maneless Lion is also said to be represented on the monuments of Upper Egypt. The Lion of Arabia has neither the courage nor the stature, nor even the beauty, of the Lion of Africa. He uses cunning rather than force; he crouches among the reeds which border the Tigris and Euphrates, and springs upon all the feeble animals which come there to quench their thirst; but he dares not attack the boar, which is very common there, and flies as soon as he perceives a man, a woman, or even a child. If he catches a sheep he makes off with his prey; but he abandons it to save himself when an Arab looks after him. If he is hunted by horsemen, which often happens, he does not defend himself unless he is wounded, and has no hope of safety by flight. In such a case he will fly on a man and tear him to pieces with his claws, for it is courage more than strength that he wants. Achmed, Pasha of Bagdad from 1724 to 1747, would have been torn by one, after breaking his lance in a hunt, if his slave Suleiman, who succeeded him in the Pashalik, had not come promptly to his succour and pierced with a blow of his yataghan the Lion already wounded by his master.

In December, 1833, Captain Walter Smee exhibited to the Zoological Society of London the skins of a Lion and Lioness killed by him in Guzerat, and distinguished from those previously known by the absence of a mane; the tail was shorter than that of the ordinary Lion, and furnished at its tip with a much larger brush or tuft; and in the tuft of the older Lion was a short horny claw or nail. The colour is fulvous; which in darker specimens has a tinge of red. A male maneless Lion, killed by Captain Smee, measured, including the tail, 8 feet $9\frac{1}{2}$ inches in length; the impression of his paw on the sand $6\frac{1}{4}$ inches across, and his height was 3 feet 6 inches. These maneless Lions are found in Guzerat, along the banks of the Sombermultee, in low, bushy-wooded plains, being driven out of the large adjoining tracts of high grass jungle by the natives annually setting fire to the grass. Here Captain Smee killed his finest specimens: they were so common in this district that he killed no fewer than eleven during a residence of about a month, yet scarcely any of the natives had seen them previously to his coming amongst them. The cattle were frequently carried off by these Lions: some natives attributed this to tigers, which, however, do not exist in this part of the country. Captain Smee could not learn that men had been attacked by these Lions: when struck by a ball they exhibited great boldness, standing as if preparing to resist their pursuers, and then going off slowly, and in a very sullen manner.

In captivity the Lioness usually turns extremely savage when she becomes a mother; and, in a state of nature, both parents guard their young with the greatest jealousy. Early in the year 1823 General Watson, then on service in Bengal, being out one morning on horseback, armed with a double-barrelled rifle, was suddenly surprised by a large male Lion, which bounded out upon him from the thick jungle, at the distance of only a few yards. He instantly fired, and the shot taking complete effect, the animal fell almost dead at his feet. No sooner had the Lion fallen than the Lioness rushed out, which the General also shot at and wounded severely, so that she retired into the thicket. Thinking that the den could not be far distant, he traced her to her retreat, and there despatched her; and in the den were found two beautiful cubs, a male and a female, apparently not more than three months old. This is a very touching narrative, even of the Lion family.

The General brought the cubs away; they were suckled by a goat and sent to England, where they arrived in September, 1823, as a present to George IV., and were lodged in the Tower. When young, Lions mew like a cat; at the age of ten or twelve months the mane begins to appear in the male; at the age of eighteen months this appendage is considerably developed, and they begin to roar. The *roar* of the adult Lion is terrific, from the larynx or upper part of the windpipe being proportionately greater than in the whale or the elephant, or any other animal. Mr. Burchell describes the roar on some occasions to resemble the noise of an earthquake; and this terrific effect is produced by the Lion laying his head upon the ground and uttering, as it were, a half-stifled roar or growl, which is conveyed along the earth.

The natural period of the Lion's life is generally supposed to be twenty or twenty-two years. Such is Buffon's limitation; but the animal will, it seems, live much longer. Pompey, the great Lion, which died in 1766, was said to have been in the Tower above seventy years; and a Lion from the river Gambia is stated to have since died in the Tower menagerie at the age of sixty-three.

There had been for ages a popular belief that the Lion lashes his sides with his tail to stimulate himself into rage; when, in 1832, there was exhibited to the Zoological Society a claw obtained from the tip of the tail of a Barbary Lion, presented to the Society's menagerie by Sir Thomas Reade. It was detected on the living animal by Mr. Bennett, and pointed out to the keeper, in whose hands it came off while he was examining it. Blumenbach quotes Homer, Lucan, and Pliny, among others who have described the Lion (erroneously) as lashing himself with his tail, when angry, to provoke his rage. None of these writers, however, advert to any peculiarity in the Lion's tail to which so extraordinary a function might, however incorrectly, be attributed. Didymus Alexandrinus, a commentator on the "Iliad," cited by Blumenbach, having found a black prickle, like a horn, among the hair of the tail, immediately conjectured that he had ascertained the true cause of the stimulus when the animal flourishes his tail in defiance of his enemies, remarking that, when punctured by this prickle, the

Lion became more irritable from the pain which it occasioned. The subject, however, appears to have slumbered till 1829, when M. Deshayes announced that he had found the prickle both of a Lion and Lioness, which had died in the French menagerie, and described it as a little nail, or horny production, adhering by its base only to the skin, and not to the last caudal vertebra. From that period Mr. Wood, the able zoologist, examined the tail of every Lion, living or dead, to which he could gain access; but in no instance had he succeeded in finding the prickle till the above specimen, which was placed in his hands within half an hour after its removal from the living animal, and while yet soft at its base, where it had been attached to the skin. Its shape was nearly straight, then slightly contracted, forming a very obtuse angle, and afterwards swelling out like the bulb of a bristle, to its termination. It was laterally flattened throughout its entire length, which did not amount to quite three-eighths of an inch, of horn colour, and nearly black at the tip. Its connexion with the skin must have been very slight, which accounts for its usual absence in stuffed as well as living specimens. This does not depend upon age, as it was found alike in the Paris Lions, of considerable size, as well as in the Zoological Society's Lions, very small and young; nor did it depend upon sex. It appears to be occasionally present in the Leopard; and, in both Lion and Leopard, it is seated at the extreme tip of the tail, and is altogether unconnected with the terminal caudal vertebra; not fitted on like a cap, but rather inserted into the skin.

The use of the prickle, however, it still remained difficult to conjecture; but that its existence was known to the ancients is proved by the Nimroud sculptures in the British Museum, in an exaggerated representation of the claw, in support of this curious fact in natural history. The existence of the claw has been proved by Mr. Bennett; and "it is no small gratification to be able now to quote in evidence of the statement of Mr. Bennett, and of his predecessor, Didymus, of Alexandria, the original and authentic document, on the authority of the veritable descendants of the renowned hunter Nimroud; which any one may read who will take the trouble to examine the sculptured slab in the British Museum."*

In the Nineveh galleries of the British Museum we also see pictured in stone the employment of the Lion, in the life of Assyria and Babylonia, three thousand years since; in the events of a succession of dynasties, recording the sieges of cities, the combats of warriors, the triumphs of Kings, the processions of victors, the chains and fetters of the vanquished. To the zoological observer these sculptures present drawings *ad naturam* of tableaux of Lions and Lion-hunts; Lions in combat, as well as in moveable dens and cages, and the ferocity of the chase; and Lions transfixed with arrows or javelins in the arena. One of the finest of these sculptures

* Bonomi; "Nineveh and its Palaces," p. 249.

is in the representation of a Lion-hunt, on a long slab that lined the principal chamber of the most ancient palace at Nimroud. The King is in his chariot, drawn by three horses, which the charioteer is urging forward to escape the attack of an infuriated Lion that has already placed its fore-paws upon the back of the chariot. At this critical moment, the Royal descendant of the mighty hunter aims a deadly shaft at the head of the roaring and wounded Lion, the position of whose tail and limbs is finely indicative of rage and fury. Behind the Lion are two of the King's attendants, fully armed, and holding their daggers and shields, ready to defend themselves in case the prey should escape the arrow of the King. Before the chariot is a wounded Lion, crawling from under the horses' feet. The cringing agony conveyed in its entire action is well contrasted with the undaunted fury of the former. In another slab we have the continuation of the same Lion-hunt, representing the triumphant return of the King from the chase. At his feet lies the Lion subdued, but not dead.

Of the pageantry of the Lion, we read, in Bell's "Travels," that the monarch of Persia had, on days of audience, two great Lions chained on each side of the passage to the state-room, led there by keepers in golden chains.

Our early English Sovereigns had a menagerie in the Tower from the reign of Henry III. (1252.) In 1370 (44 Edward III.) are entries of payments made to "the Keeper of the King's Lions and

Leopards" there, at the rate of 6*d*. a-day for his wages, and 6*d*. a-day for each beast. The number of beasts varied from four to seven. Two young Lions are specially mentioned; and "a Lion lately sent by the Lord the Prince, from Germany to England, to our Lord the King." And we read, in Lord Burghley's "Diary," 1586, of the grant of the keeping of the Lions in the Tower, with "the Fine of 12*d*. per diem, and 6*d*. for the Meat of those Lions." The first menagerie-building was the Lion Tower, to which was added a semicircular inclosure, where Lions and Bears were baited with dogs, with which James I. and his court were much delighted. A Lion was named after the reigning King; and it was popularly believed that " when the King dies, the Lion of that name dies after him." The last of the Tower animals were transferred to the Zoological Society's menagerie, in the Regent's-park, in 1834. The Tower menagerie is well described in a handsome volume, with woodcut portraits, by William Harvey.

The punishment of being *thrown to Lions* is stated as common among the Romans of the first century; and numerous tales are extant, in which the fierce animals became meek and lamb-like before the holy virgins of the Church. This, indeed, is the origin of the superstition, nowhere more beautifully expressed than in Lord Byron's " Siege of Corinth ":—

> " 'Tis said that a Lion will turn and flee
> From a maid in the pride of her purity."

Every wild beast show almost has its tame Lion, with which the keeper takes the greatest liberties; liberties which the beast will suffer, generally speaking, from none but him. Major Smith relates that he had seen the keeper of a Lioness stand upon the beast, drag her round the cage by her tail, open her jaws, and thrust his head between her teeth. Another keeper, at New York, had provided himself with a fur cap, the novelty of which attracted the notice of the Lion, which, making a sudden grapple, tore the cap off his head as he passed the cage; but, perceiving that the keeper was the person whose head he had thus uncovered, he immediately laid the cap down. Wombwell, in his menagerie, had a fine Lion, Nero, that allowed *strangers* to enter his den, and even put their heads within his jaws. This tameness is not, however, to be trusted, since the natural ferocity of some Lions is never safely subdued. Lions which have been sometimes familiar, have, on other occasions, been known to kill their keepers, and dart at those who have incautiously approached too near their cage. All these exhibitions have been entirely eclipsed by the feats of Van Amburgh, in his exercise of complete control over Lions. The melancholy fate of "the Lion Queen," however, tells of the fatal result of her confidence. The Lion-killing feats of Captain Gordon Cumming had a more legitimate object in view—to render us more familiar with the zoological character of the Lion.

Colonization has scarcely yet extirpated the Lion in Algeria, where the French colonists make fine sport of "the King of the Beasts." M. Jules Gerard, a Nimroud in his way, has been noted for his Lion-killing feats. We read of his tracking a large old Lion in the Smauls country, one hundred leagues in ten days, without catching a glimpse of anything but his foot-prints. At length, accompanied by a native of the country and a spahi, Gerard took up his quarters at the foot of a tree upon the path which the old Lion had taken. It was moonlight, and Gerard made out two Lions sitting about one hundred paces off, and exactly in the shadow of the tree. The Arab lay snoring ten paces off, in the full light of the moon, and had, doubtless, attracted the attention of the Lions. Gerard expressly forbade the spahi to wake the Arab. Our Lion-hunter then got up the hill to reconnoitre; the boldest of the Lions came up to within ten paces of Gerard, and fifteen of the Arab: the Lion's eye was fixed on the latter, and the second Lion placed himself on a level with, and four or five paces from, the first. They proved to be both full-grown Lionesses. Gerard took aim at the first as she came rolling and roaring down to the foot of the tree. The Arab was scarcely awakened, when a second ball stretched the Lioness dead upon the spot. Gerard then looked out for the second Lioness, who was standing up within fifteen paces, looking around her. He fired, and she fell down roaring, and disappeared in a field of maize; she fell, but was still alive. Next morning at daybreak, at the spot where the Lioness had fallen, were blood marks, denoting her track in the direction of a wood. After sending off the dead Lioness, Gerard returned to his post of the preceding night. A little after sunset the Lion roared in his lair, and continued roaring all night. Convinced that the wounded Lioness was there, Gerard sent two Arabs to explore the cover, but they durst not. He next evening reached the lair, taking with him a goat, which he left with the Arabs: the Lioness appeared, Gerard fired, and she fell without a struggle; she was believed dead, but she got up again as though nothing was the matter, and showed all her teeth. One of the Arabs, within six paces of her, seeing her get up, clung to the lower branches of a tree and disappeared like a squirrel. The Lioness fell dead at the foot of the tree, a second bullet piercing her heart: the first had passed out of the nape of the neck without breaking the skull-bone.

The Lions presented by Lord Prudhoe to the British Museum are the best sculptured representations of the animal in this country. Although the Lion is our national hieroglyphic, and there are many statues of him, yet not one among them all appears without a defect, which makes our representations of him belong to the class *canis* instead of *felis*, a fault not found in any Egyptian sculpture.*

* Bonomi; "Proc. Royal Soc., Literature."

"Behold the fowls of the air : for they sow not, neither do they reap, nor gather into barns ; yet your Heavenly Father feedeth them."—Matthew vi. 26.

> "Free tenants of land, air, and ocean,
> Their forms all symmetry, their motions grace ;
> In plumage delicate and beautiful,
> Thick without burthen, close as fishes' scales,
> Or loose as full-blown poppies on the gales ;
> With wings that seem as they'd a soul within them,
> They bear their owners with such sweet enchantment."
> *James Montgomery.*

BIRDS, as regards structure, are perhaps the most perfectly endowed, as they are certainly the most beautiful and interesting, of all the lower animals. In Birds there is an admirable mechanism and adaptation both for gliding in the air and swimming in the water. They surpass all other animals in the faculty of continuing their motion without resting, as well as in its rapidity. The fleetest courser can scarcely ever run more than a mile in a minute, nor support that speed beyond five or six such exertions. But the

joyous Swallow does this tenfold for pleasure. In his usual way he flies at the rate of one mile in a minute; and Wilson, the ornithologist, ascertained that the Swallow is so engaged for ten hours every day. So can the Blue-bird of America, for a space of 600 miles. Our Carrier-pigeons move with half that celerity: one flew from Liskeard to London, 220 miles, in six hours. The Golden Eagle is supposed to dart through the fiercest storm at the rate of 160 miles an hour; but one of our smallest Birds, the Swift, can even quadruple the most excited quickness of the race-horse for a distance. Spallanzani thought that the little Swift travelled at the rate of 250 miles an hour.

Inquiries into the phenomena of the flight of Birds would lead us far beyond our limits. The subject is beset with error. Thus, we read:—"Every one has remarked the manner in which Birds of prey float, as it were, without any effort, and with steady expanded wings, at great heights in the atmosphere. This they are enabled to do from the quantity of air contained in the air-cells of their bodies, which air being taken in at a low level in the atmosphere, of course rarefies and expands as the Bird ascends into higher regions. Their rapidity of descent must be accomplished by the sudden expulsion of this air, aided by their muscular efforts."

Now, Dr. Crisp has read to the Zoological Society a paper "On the Presence or Absence of Air in the Bones of Birds," for the purpose of showing the prevailing error upon the subject—viz., "that the bones of the Bird are filled with air." Of fifty-two British Birds recently dissected by him, only one, the Sparrow-hawk, had the bones generally perforated for the admission of air. In thirteen others, the humeri only were hollow, and among these were several Birds of short flight. In the remaining thirty-eight, neither the *humeri* nor *femora* contained air, although in this list were several Birds of passage and of rapid flight—Dr. Crisp's conclusion being, that the majority of British Birds have no air in their bones, and that, with the exception of the Falcons, but very few British Birds have hollow femora.

Mr. Gould records a most remarkable instance of rapid and sustained flight, which he witnessed on his return from North America, whither he had proceeded for the purpose of studying the habits and manners of the species of *Trochilus* (Humming Bird), frequenting that portion of America. Having remarked that he arrived just prior to the period of the migration of this Bird from Mexico to the north, and had ample opportunities for observing it in a state of nature, he noticed that its actions were very peculiar, and quite different from those of all other birds: the flight is performed by a motion of the wings so rapid as to be almost imperceptible; indeed, the muscular power of this little creature appears to be very great in every respect, as, independently of its rapid and sustained flight, it grasps the small twigs, flowers, &c., upon which it alights with the utmost tenacity. It appears to be most active in the morning and evening, and to pass the middle of the day in a state of sleepy torpor. Occasionally it occurs in such numbers that fifty or sixty birds may be seen in a single tree. When captured it so speedily becomes tame that it will feed from the hand or mouth within half an hour. Mr. Gould having been successful in keeping a Humming-Bird alive in a gauze bag attached to his breast button for three days, during which it readily fed from a bottle filled with a syrup of brown sugar and water, he determined to make an attempt to bring some living examples to England, in which he succeeded; but unfortunately they did not long survive their arrival.

The adaptation of colour in Birds to their haunts strikingly tends to their preservation. The small Birds which frequent hedges have backs of a brownish or brownish-green hue; and their bellies are generally whitish, or light-coloured, so as to harmonise with the sky. Thus, they become less visible to the hawk or cat that passes above or below them. The wayfarer across the fields also treads upon the Skylark before he sees it warbling to heaven's gate. The Goldfinch or Thistlefinch passes much of its time among flowers, and is vividly coloured accordingly. The Partridge can hardly be distinguished from the fallow or stubble among which it crouches; and it is considered an accomplishment among sportsmen to have a good eye for finding a Hare sitting. In northern countries the winter dress of the Hares and Ptarmigans is white, to prevent detection among the snows of those inclement regions.

The Song of Birds is popularly explained by the author of a work, entitled, "The Music of Nature," in which he illustrates the vocal machinery of Birds as follows :—" It is difficult to account for so small a creature as a Bird making a tone as loud as some animal a thousand times its size; but a recent discovery shows that in birds the lungs have several openings communicating with corresponding air-bags or cells, which fill the whole cavity of the body from the neck downward, and into which the air passes and repasses in the progress of breathing. This is not all. The very bones are hollow, from which air-pipes are conveyed to the most solid parts of the body, even into the quills and feathers. The air being rarefied by the heat of their body, adds to their levity. By forcing the air out of their body, they can dart down from the greatest heights with astonishing velocity. No doubt the same machinery forms the basis of their vocal powers, and at once resolves the mystery into a natural ordering of parts." This is a very pretty story; but, unfortunately, it is not correct, as already shown.

A correspondent of the "Athenæum," writing in 1866, says :—" He would be a bold man who should say that Birds have no delight in their own songs. I have been led to conclude from experiments which I have made, and from other observations, that certain animals, especially Birds, have not only an ear for fine sounds, but also a preference for the things they see out of respect to fine colours or other pleasing external features. It is chiefly among Birds, when we consider the case of animals, that a taste for ornament and for glittering objects, often very startling and human-like, is to be found. The habits of the Pheasant, Peacock, Turkey, Bird of Paradise, several Birds of the Pigeon and Crow kind, and certain Singing Birds, are evidence. The Australian Satin Bower-Bird is the most remarkable of that class which exhibit taste for beauty or for glittering objects out of themselves—that is, beauty not directly personal; collecting, in fact, little museums of shells, gaudy feathers, shining glass, or bits of coloured cloth or pottery. It will be found with many Birds that fine plumes, a mirror, and an admirer, are not altogether objects devoid of interest.

"Another consideration leading me to the same conclusion, is the fact, that beauty in animals is placed on prominent parts, or on parts which by erection or expansion are easily, and at the pairing season, frequently rendered prominent, such as a crest or tail. A spangle of ruby or emerald does not exist, for instance, on the side under the wing, which is seldom raised, of our domestic poultry. Such jewels are hung where man himself wears his, on the face and forehead, or court attention, like our own crowns, trains, shoulder-knots, breast-knots, painted cheeks, or jewelled ears. I cannot account for the existence of these gaudy ornaments to please man, for nowhere are they more gorgeous than in Birds which live in the depth of the tropical forest, where man is rarely a visitor; I cannot account for them on the principle that they do good to their possessors in the battle for life, because they rather render them conspicuous to their enemies, or coveted by man." But the beauty of these beings glows most brightly at the season of their pairing, and the selection of their mates.

Baron von Tschudi, the Swiss naturalist, has shown the important services of Birds in the destruction of insects. Without Birds, no agriculture or vegetation would be possible. They accomplish in a few months the profitable work of destruction which millions of human hands could not do half so well in as many years; and the sage, therefore, blamed in very severe terms the foolish practice of shooting and destroying Birds, which prevails more especially in Italy, recommending, on the contrary, the process of alluring Birds into gardens and cornfields. Among the most deserving Birds he counts Swallows, Finches, Titmice, Redtails, &c. The naturalist then cites numerous instances in support of his assertion. In a flower-garden of one of his neighbours three rose-trees had been suddenly covered with about 2,000 tree-lice. At his recommendation a Marsh-Titmouse was located in the garden, which in a few hours consumed the whole brood, and left the roses perfectly clean. A Redtail in a room was observed to catch about 900 flies in an hour. A couple of Night-Swallows have been known to destroy a whole swarm of gnats in fifteen minutes. A pair of Golden-crested Wrens carry insects as food to their nestlings upon an average thirty-six times in an hour. For the protection of orchards and woods Titmice are of invaluable service. They consume, in particular, the eggs of the dangerous pine-spiders. One single female of such spiders frequently lays from 600 to 800 eggs twice in the summer season, while a Titmouse with her young ones consume daily several thousands of them. Wrens, Nuthatches, and Woodpeckers often dexterously fetch from the crevices of tree-bark numbers of insects for their nestlings.

Yet, profitless and wanton Bird-murder is common. The cliffs on the coasts of these islands are the resort of numerous kinds of Sea-Fowl, and these Fowl, we are told, are slaughtered by thousands, not merely for the sake of their feathers, but actually for the mere savage pleasure of killing. What speculation can enter into such a proceeding it may puzzle the reader to imagine; but it seems that the wing feathers of the poor White Gull are now inquired for in the plume-trade, and we are actually told of an order given by a single house for 10,000 of these unhappy Birds. When these facts were stated at the Meeting of the British Association, in August, 1868, at Norwich, a lady stood up boldly in defence of her sex, and declared that they sinned only through ignorance, and would never willingly wear the feathers of a Bird destroyed in the act of feeding its young. That part of the case, therefore, ought to be now in safe hands. In the Isle of Man a law has been passed, called the "Seagull Preservation Act," protecting these Birds by heavy penalties, on the ground of their utility in removing fish offal and guiding fishermen to shoals of fish. At a certain point of our shores a similar protection has been established. A visitor to the South Stack Lighthouse, on the coast of Anglesey, may see prodigious numbers of Sea-Fowl as tame as complete safety can make them. It has been ascertained that in thick weather, when neither light can be distinguished nor signal seen, the incessant scream of these Birds gives the best of all warnings to the mariner of the vicinity of the rock. The noise they make can be heard at a greater distance than the tolling of the great bell; and so valuable was this danger-signal considered, that an order from the Trinity House forbad even the firing of the warning gun, lest the colony of the Sea-Fowl should be disturbed. The signals of the bell and the cannon might be neglected or overpowered, but the Birds were always there and always audible.

It is inferred that Birds possess some notion of power, and of cause and effect, from the various actions which they perform. "Thus," relates Dr. Fleming, "we have seen the Hooded Crow in Zetland, when feeding on small shell-fish, able to break some of the tenderer kinds by means of its bill, aided in some cases by beating them against a stone; but, as some of the larger shells, such as the buckie and the welk, cannot be broken by such means, the Crow employs another method, by which, in consequence of applying foreign power, it accomplishes its object. Seizing the shell with its claws, it mounts up into the air, and then loosing its hold, causes the shell to fall among stones (in preference to the sand, the water, or the soil on the ground), that it may be broken, and give easier access to the contained animal. Should the first attempt fail, a second or third is tried, with this difference, that the Crow rises higher in the air, in order to increase the power of the fall, and more effectually remove the barrier to the contained morsel. On such occasions we have seen a strong Bird remain an apparently inattentive spectator of the process of breaking the shell, but coming to the spot with astonishing keenness when the efforts of its neighbour had been successful, in order to share the spoil. Pennant mentions similar operations performed by Crows on mussels."

The brain of Birds is, in general, large in proportion to the size of the body, and the instinctive powers are very perfect. A few kinds are rather dull and stupid; but the Parrot, Magpie, Raven, and many others, show great vivacity and quickness of intellect. The Raven has a great deal of humour in him. One, a most amusing and mischievous creature, would get into a well-stocked flower-garden, go to the beds where the gardener had sowed a great variety of seeds, with sticks put in the ground with labels, and then he would amuse himself with pulling up every stick, and laying them in heaps of ten or twelve on the path. This used to irritate the old gardener, who drove him away. The Raven knew that he ought not to do it, or he would not have done it. He would soon return to his mischief, and when the gardener again chased him (the old man could not run very fast), the Raven would just keep clear of the rake or the hoe in his hand, dancing before him, and singing as plainly as a Raven could, "Tol de rol de rol! tol de rol de rol!" with all kinds of mimicking gestures.

The signal of danger among Birds seems to be of universal comprehension; because the instant it is uttered we hear the whole flock, though composed of various species, repeat a separate moan, and away they all scuttle into the bushes for safety. The sentinel Birds give the signal, but in some cases they are deceived by false appearances. Dr. Edmonstone, in his "View of the Zetland Isles," relates a very striking illustration of the neglect of the sentinel, in his remarks on the Shag. "Great numbers of this species of the Cormorant are sometimes taken during the night, while asleep on the rocks of easy access; but before they commit themselves to sleep, one or two of the number are appointed to watch. Until these sentinels are secured, it is impossible to make a successful impression on the whole body; to surprise them is, therefore, the first object. With this view, the leader of the expedition creeps cautiously and imperceptibly along the rock, until he gets within a short distance of the watch. He then dips a worsted glove into the sea, and gently throws water in the face of the guard. The unsuspecting Bird, either disliking the impression, or fancying, from what he considers to be a disagreeable state of the weather, that all is quiet and safe, puts his head under his wing and soon falls asleep. His neck is then immediately broken, and the party dispatch as many as they choose."

Addison was a true lover of nature, which he shows in two letters written by him to the Earl of Warwick (afterwards his son-in-law), when that nobleman was very young. "My dear Lord," he writes, "I have employed the whole neighbourhood in looking after Birds'-nests, and not altogether without success. My man found one last night, but it proved a hen's, with fifteen eggs in it, covered with an old broody Duck, which may satisfy your Lordship's curiosity a little; though I am afraid the eggs will be of little use to us. This morning I have news brought me of a nest that has abundance of little eggs, streaked with red and blue veins, that, by the description they give me, must make a very beautiful figure in a string. My neighbours are very much divided in their opinions upon them: some say they are a Skylark's; others will have them to be a Canary-Bird's; but I am much mistaken in the colour and turn of the eggs if they are not full of Tomtit's." Again, Addison writes:—"Since I am so near your Lordship, methinks, after having passed the day amid more severe studies, you may often take a trip hither and relax yourself with these little curiosities of nature. I assure you no less a man than Cicero commends the two great friends of his age, Scipio and Lælius, for entertaining themselves at their country-house, which stood on the sea-shore, with picking up cockle-shells, and looking after Birds'-nests."

In another letter Addison writes:—"The business of this is to invite you to a concert of music which I have found out in a neighbouring wood. It begins precisely at six in the evening, and consists of a Blackbird, a Thrush, a Robin-Redbreast, and a Bullfinch. There is a Lark, that, by way of overture, sings and mounts till she is almost out of hearing; and afterwards, falling down leisurely, drops to the ground as soon as she has ended her song. The whole is concluded by a Nightingale, that has a much better voice than Mrs. Tofts, and something of the Italian manner in her divisions. If your Lordship will honour me with your company, I will promise to entertain you with much better music, and more agreeable scenes, than you ever met with at the Opera; and will conclude with a charming description of a Nightingale out of our friend Virgil:—

"'So close, in poplar shades, her children gone,
The mother Nightingale laments alone;
Whose nest some prying churl had found, and thence
By stealth convey'd the unfeathered innocence:
But she supplies the night with mournful strains,
And melancholy music fills the plains.'"

BIRDS' EGGS AND NESTS.

THE Eggs of Birds are variously tinted and mottled, and hence they become objects of interest to the collector. In this diversity of colour nature has, doubtless, some final object in view; and though not in every instance, yet in many, we can certainly see a design in the adaptation of the colours to the purpose of concealment, according to the habits of the various classes of Birds. Thus, as a general rule, the Eggs of Birds which have their nests in dark holes, or which construct nests that almost completely exclude the light, are white; as is also the case with those Birds that constantly sit on their Eggs, or leave them only for a short time during the night. Eggs of a light blue or light green tint will also be found in nests that are otherwise well concealed; while, on the other hand, a great proportion of those nests that are in exposed situations have Eggs varying in tints and spots in a remarkable degree, corresponding with the colours of external objects in their immediate neighbourhood.

Thus, a dull green colour is common in most gallinaceous Birds that form their nests in grass, and in aquatic Birds among green hedges; a bright green colour is prevalent among Birds that nestle among trees and bushes; and a brown mottled colour is found in those Eggs that are deposited among furze, heath, shingle, and grey rocks and stones.

Birds'-nesting, we need hardly remark, is a favourite pursuit of boyhood; but, in some cases, its attractions have induced young persons to take up more important branches of natural history, or the collection, systematic arrangement, and comparison of Birds' Eggs, which is, in scientific study, termed Oology; and as the study of Birds cannot be considered complete until they are known in every stage, it forms a branch of Ornithology. In this case Birds'-nesting has an useful object; but many persons are content to acquire collections of Eggs without troubling themselves about the Birds which have laid them.

The late Mr. John Wolley, M.A., was one of the leading authorities upon the subject of European Ornithology, and was one of a number of University men, who, about twelve years ago, established the ornithological journal called "The Ibis," and who visited far-distant and unexplored regions, where they might hope to discover strange Birds and unknown Eggs. For several years Algiers and Tunis were their favourite resorts, and the meeting-places of many of our rarer Birds were hunted up in these countries, even so far as the Desert of the Great Sahara. Others preferred the New World as the scene of their labours, and collected long series of specimens in the highland of Guatemala, and the tropical forests of Belize. Mr. Wolley, however, confined his attention principally to the northern parts of Europe—that region being the breeding-quarters of a large number of Birds which are only known in this country as winter visitants. In order to be at his collecting-station at Muonioniska, on the frontier of Finnish Lapland, at the earliest commencement of the breeding-season, Mr. Wolley frequently passed the whole winter in that remote region. But the rigour of the climate under the Arctic Circle contributed to bring on a malady which terminated fatally in November, 1859.

Upon the decease of Mr. Wolley, his large collection of Birds' Eggs, in accordance with his last wishes, became the property of his friend, Mr. Alfred Newton, who is publishing a Catalogue of Mr. Wolley's Egg Cabinet, with notes from the deceased naturalist's journals. The first part contains the Eggs of Birds of Prey (*Accipitres*), recognisable at once by their strongly-hooked bill, formed to assist them in tearing their prey, and their large feet and sharpened claws, which aid them to grasp it. They are divisible into two very distinct groups—the diurnal Birds of Prey, consisting of the Hawks, Vultures, and Eagles; and the nocturnal Birds of Prey, or Owls. In the latter the Eggs are invariably colourless; in the former they are often strongly marked, and present some of the most beautiful objects in the whole series of Birds' Eggs.

In the most recently published list of European birds fifty-two species of birds of prey are given as occurring more or less frequently within the limits of our continent. Of the three generally-recognised species of European Vultures two are well represented, as regards their eggs, in the Wolleyan series. A few years ago the nesting of all these birds was utterly unknown to naturalists, and it was mainly through the exertions of Mr. Wolley and his friends that specimens first reached our collectors' cabinets. Here were found both the Egyptian Vulture and the Griffon breeding abundantly in the Eastern Atlas in 1857; and the eyries of these birds have since been visited by other collectors in the same country. The Eggs of the former of these Vultures are remarkable for their deep and rich coloration. The productions of the Griffon are not nearly so handsome, and are occasionally altogether destitute of markings. Of the Eagles of Europe the series of Eggs is very full, especially of the two well-known British species—the Golden Eagle and Sea Eagle. The Golden or Mountain Eagle is even now-a-days much more common in the remote parts of the British islands than is usually supposed to be the case. In 1852 Mr. Wolley was acquainted with five nests of this bird in various parts of Scotland, and there were undoubtedly at least as many more of which he did not learn the particulars. The eyrie is usually placed in some mountainous district, on the ledge of some "warm-looking" rock, well clothed with vegetation, and often by no means wild or exposed. Not unfrequently, under proper guidance, one can walk into the nest almost without climbing. Mr. Newton gives a very entertaining account of the taking of a pair of eggs from a nest in Argyllshire in 1861, where this seems to have been the case. In the whole ascent there was only one "ticklish place," where it was necessary to go sideways on a narrow ledge round some rocks. The Sea Eagle, on the other hand, generally breeds on the high cliffs upon the coast, often selecting the most inaccessible position for its eyrie. Sometimes, however, it will choose an island in the middle of an inland loch, and in such case places its nest upon the ground or in a tree.

Mr. Wolley's well-written notes of his adventures in quest of both these Eagles, as also those relating to the other rapacious birds, will be read with much interest; as will also the details concerning the nesting-habits of many of the rarer species of European birds, several of which, such as the Rough-legged Buzzard and the Lapp Owl, were first tracked to their breeding-quarters in the remotest wilds of Scandinavia by this indefatigable naturalist.*

Of large Eggs we are most familiar with those of the Ostrich, of which Mr. Burchell, when in Africa,

* Abridged from the "Saturday Review."

found twenty-five Eggs in a hollow scratched in the sand, six feet in diameter, surrounded by a trench, but without grass, leaves, or sticks, as in the nests of other birds. In the trench were nine more Eggs, intended, as the Hottentots observed, as the first food of the twenty-five young Ostriches. Between sixty and seventy Eggs have been found in one nest; each is equal to twenty-four Eggs of the domestic hen, and holds five pints and a quarter of liquid. The shells are dirty white. The Hottentots string them together as belts, or garlands, and they are frequently mounted as cups. One Ostrich Egg is a sufficient meal for three persons. The Egg is cooked over the fire without either pot or water, the shell answering the purpose of the first, and the liquid nature of its contents that of the other.

Less familiar to the reader are the gigantic Eggs of the Epyornis, a bird which formerly lived in Madagascar. One of these Eggs contains the substance of 140 hens' Eggs. Mr. Geoffroy St. Hilaire describes some portions of an Egg of the Epyornis which show the Egg to have been of such a size as to be capable of containing about ten English quarts; that in the Museum of the Jardin des Plantes can only contain $8\frac{3}{4}$ quarts. Mr. Strickland, in some notices of the Dodo and its kindred, published in 1849, says that in the previous year a Mr. Dumarele, a French merchant at Bourbon, saw at Port Leven, Madagascar, an enormous Egg which held "*thirteen wine quart bottles of fluid.*" The natives stated that the Egg was found in the jungle, and "that such Eggs were *very, very rarely* met with."

A word or two about the nests of such gigantic birds. Captain Cook found, on an island near the north-east coast of New Holland, a nest "of a most enormous size. It was built with sticks upon the ground, and was no less than six-and-twenty feet in circumference, and two feet eight inches high." (Kerr's "Collection of Voyages and Travels," xiii., 318.) Captain Flinders found two similar nests on the south coast of New Holland, in King George's Bay. In his "Voyage," &c., London, 1818, he says, "They were built upon the ground, from which they rose above two feet, and were of vast circumference and great interior capacity; the branches of trees and other matter of which each nest was composed being enough to fill a cart."

Among the varieties of Birds'-nests are some very curious homes, of which we have but space to notice a few. The pendulous nest of the Indian Baya-bird is usually formed of the fibres of the palmyra, the cocoa-nut palm, and wild date of India, sometimes mixed with grass, neatly interlaced, and very strongly made. It consists of only one circular chamber, with a long tubular passage leading to it, and is suspended from a tree, preferred if overhanging water. The natives of India say the Baya lights up its nest with fire-flies. The bird lays from four to six white eggs. Bayas are of a very social disposition: numbers build on the same tree, or neighbouring trees, and singing in concert during the breeding season. The Baya is very docile, and taught to fly off the finger and return again; to dart after a ring or small coin, dropped into a deep well, and catch it before it reaches the water; to fetch and carry, and perform similar tricks.

The nest of the brilliant Golden-banded Oriole is a hammock of twisted fibrous substances, and is suspended in a low shrub, so as to swing to the breeze. The twine-like fibres of which it is woven are the filaments of the gigantic palm. The threads break away from the leaf, and hang like fringe to the magnificent foliage.

The Tailor-birds are the best nest-builders of all the feathered tribes. They interweave their nests between the twigs and branches of shrubs, or suspend the nests from them; and some of these birds have exercised arts from the creation which man has found of the greatest benefit to him since he discovered them. These birds, indeed, may be called the inventors of the several arts of the weaver, the sempstress, and the tailor; whence some of them have been denominated Weaver and Tailor Birds. The nests of the latter are, however, most remarkable. India produces several species of Tailor-birds that sew together leaves for the protection of their eggs and nestlings from the voracity of serpents and apes. They generally select the end of a branch or twig, and sew with cotton, thread, and fibres. Colonel Sykes has seen some in which the thread was literally knotted at the end. The inside of these nests is lined usually with down and cotton.

Tailor-birds are not confined to India or tropical countries. Italy can boast a species which exercises the same art. Mr. Gould has a specimen of this bird in his possession, and the Zoological Society have a nest in their Museum. This little bird, a species of the genus *sylvia*, in summer and autumn frequents marshes; but in the spring it seeks the meadows and cornfields, in which, at that season, the marshes being bare of the sedges which cover them in summer, it is compelled to construct its nest in tussocks of grass on the brinks of ditches; but the leaves of these being weak, easily split, so that it is difficult for our little sempstresses to unite them, and so form the skeleton of the fabric. From this and other circumstances, the spring nests of these birds differ so widely from those made in the autumn that it seems next to impossible that both should be the work of the same artisan. The latter are constructed in a thick bunch of sedge or reed: they are shaped like a pear, being dilated below and narrow above, so as to leave an aperture sufficient for the ingress and egress of the bird. The greatest horizontal diameter of the nest is about two inches and a half, and the vertical is five inches.

The most wonderful thing in the construction of these nests is the method to which the little bird has recourse to keep united the living leaves of which it is composed. The sole in the weaving, more or less delicate, of the materials, forms the principle adopted by other birds to bind together the walls of their nests; but this sylvia is no weaver, for the leaves of the sedges or reeds are united by real stitches. In the edge of each leaf she makes, probably with her beak, minute apertures, through which she contrives to pass, perhaps by means of the same organ, one or more cords formed of spiders' web, particularly that of their egg-pouches. Those threads are not very long, and are sufficient to pass two or three times from one leaf to another. They are of unequal thickness, and have knots here and there, which, in some places, divide into two or three branches.

This is the manner in which the exterior of the nest is formed: the interior consists mainly of down, chiefly from plants, a little spiders' web being intermixed, which helps to keep the other substances together. The upper part and sides of the nest, that is, the external and internal, are in immediate contact; but in the lower part a greater space intervenes, filled with the slender foliage of grasses, and other materials, which render soft and warm the bed on which the eggs are to repose. This little bird feeds on insects. Its flight is rectilinear, but consists of many curves, with the concavity upwards. These curves equal in number the strokes of the wing, and at every stroke its whistle is heard, the intervals of which correspond with the rapidity of its flight.

The Australian Bower-bird, as its name implies, builds its nest like an arbour or bower, with twigs: in the British Museum are two specimens, each decorated—one with bones and fresh-water shells, and the other with feathers and land-shells; remarkable instances of taste for ornament already referred to in a preceding page. The Satin or Bower-bird is described by settlers in Australia as "a very troublesome rascal," which besets gardens; if once allowed to make a lodgment there it is very troublesome to get rid of him; he signalizes his arrival by pulling up, in his restless fussy way, everything in the garden that he can tug out of the ground, even to the little sticks to mark the site of seeds. A settler had formed a garden in the bush; there was no enclosure of the kind for miles in any direction: a flock of Bower-birds came; he got his gun and shot two or three; the flock went off, and he never saw another bird of the kind.

The Cape Swallows build nests which show extraordinary instinct allied to reason. A pair of these built their nest on the outside of a house at Cape Town against the angle formed by the wall and the board which supported the eaves. The whole of this nest was covered in, and it was furnished with a long neck or passage, through which the birds passed in and out. It resembled a longitudinal section of a Florence oil flask. This nest having crumbled away after the young birds had quitted it, the same pair, or another of the same species, built on the old foundation again. But this time an improvement was observable in the plan of it that can hardly be referred to the dictates of mere instinct. The body of the nest was of the same shape as before, but instead of a single passage it was furnished with one at each side, running along the angle of the roof; and on watching the birds, they were seen invariably to go in at one passage and come out at the other. Besides saving themselves the trouble of turning in the nest and disturbing, perhaps, its interior arrangement, they were guarded by this contrivance against a surprise by serpents, which frequently creep up along the wall, or descend from the thatch, and devour both the mother and her brood.

Dr. Livingstone relates a very curious instance of "Bird Confinement" under very strange circumstances. In passing through Mopane country, in South Africa, his men caught a great number of the birds called *Korwé* in their breeding-places, which were holes in the mopane trees. They passed the nest of a Korwé just ready for the female to enter; the orifice was plastered on both sides, but a space was left of a heart shape, and exactly the size of the bird's body. The hole in the tree was in every case found to be prolonged some distance upwards above the opening, and thither the Korwé always fled to escape being caught. In another nest that was found, one white egg, much like that of a pigeon, was laid, and the bird dropped another when captured: she had four besides in the ovarium. Dr. Livingstone first saw this bird at Kolenbeng in the forest: he saw a slit only, about half an inch wide and three or four inches long, in a slight hollow of a tree; a native broke the clay which surrounded the slit, put his arm into the hole, and brought out a red-beaked Hornbill, which he killed. He told Dr. Livingstone that when the female enters her nest she submits to a real confinement. The male plasters up the entrance, leaving only a narrow slit by which to feed his mate, and which exactly suits the form of his beak. The female makes a nest of her own feathers, lays her eggs, hatches them, and remains with the young till they are fully fledged. During all this time, which is stated to be two or three months, the male continues to feed her and the young family. The prisoner generally becomes quite fat, and is esteemed a very dainty morsel by the natives; while the poor slave of a husband gets so lean that, on the sudden lowering of the temperature, which sometimes happens after a fall of rain, he is benumbed, falls down, and dies.

Dr. Livingstone, on passing the same tree at Kolenbeng about eight days afterwards, found the hole plastered up again, as if, in the short time that had elapsed, the disconsolate bird-husband had procured another wife. Dr. L. saw a nest with the plastering not quite finished, and others completed; he also received elsewhere, besides Kolobeng, the same account that the bird comes forth when the young are fully-fledged, at the period when the corn is ripe; indeed, her appearance abroad with her young is one of the signs they have for knowing when it ought to be so: the time is between two and three months. She is said sometimes to hatch two eggs, and, when the young of these are full-fledged, the other two are just out of the egg-shells: she then leaves the nest with the two elder, the orifice is again plastered up, and both male and female attend to the wants of the young.

There is a specimen of a nest in the Gardens of the Zoological Society, which merits description, besides that of the Bower-bird. Such is the nest of the Brush Turkey, which appears more like a small haystack than an ordinary nest, and the methodical manner in which it is constructed is thus described:—Tracing a circle of considerable radius, the birds begin to travel round it, continually grasping with their huge feet the leaves and grasses and dead twigs which are lying about, and flinging them inwards towards the centre. Each time that they complete their round, they narrow their circle, so that in a short time they clear away a circular belt, having in its centre a low irregular mass. By repeating the same process, however, they decrease the diameter of the mound as they increase its height, and at last a large and rudely conical mound is formed.

In this nest as many as a bushel of eggs are deposited, at regular intervals, long end downwards. The leaves form a fermenting mass, which relieves the mother of the necessity of setting upon them. The male, however, has to regulate the temperature of the mass, which would otherwise get too hot. This he does by making a central ventilating shaft, which carries off the superfluous heat; and, lest the temperature should fall too low, he is constantly engaged in covering and uncovering the eggs in order to hit the exact temperature to be applied until the egg is warmed into life.

THE EPICURE'S ORTOLAN.

WE have allotted this bird to the epicure, because it is rarely heard of but in association with his luxurious table. Mr. Beckford describes the Ortolans among the delicacies which he saw in the kitchen of the monastery of Batalha as "lumps of celestial fatness."

Ortolan is the French and English names for a species of *Fringillidæ* (Finches). It is the *Hortulanus* of Gesner and other naturalists; *Miliaria pinguescens* of Frisch; *Emberiza Hortulana* of Linnæus; *Ortolano* of the Italians generally; *Tordino Berluccio* of the Venetians; *Garton Ammer* and *Fetammer* of the Germans; and *Gerste Keneu* of the Netherlanders. This wide dispersion on the Continent bespeaks the pet character of the bird. Montagu terms it the Green-headed Bunting.

The French have a fanciful derivation of the name: they say it is from the Italian word for gardener, which is from the Latin *hortus*, garden; because, according to Menage, in Italy, where the bird is common, it is quite at home in the hedges of gardens.

The male bird has the throat, circle round the eyes, and a narrow band springing from the angle of the bill, yellow; head and neck grey, with a tinge of olive, and small brown spots; feathers black, edged with red; breast, belly, and abdomen, reddish grey, the feathers terminated with ash-colour; tail blackish, two external feathers, in part white; length rather more than six inches. There are, also, varieties marked white, green, blackish, and entirely black. The nest, which is constructed of fibres of plants and leaves, is frequently found on the ground in corn-fields, and sometimes in hedges and bushes.

The Ortolan is not famed for its song, which is, however, soft and sweet. Like the Nightingale, to which it has other points of resemblance, the Ortolan sings after, as well as before sunset. It was this bird that Varro, the lyric poet, called his companion by night and day.

The south of Europe may be considered the summer and autumnal head-quarters of the Ortolan, though it is a summer visitor in the central and northern parts. In Italy it is said to be common by Temminck and others. The Prince of Musignano states it to be found in the Sabine mountains; adding that it rarely flies in the plains of Rome, but is frequent in Tuscany. Lapland, Russia, Denmark, Sweden, and Norway, are among the countries visited by it. In the British Isles it seems only entitled to rank as an autumnal visitor, but it may occur more frequently than is generally supposed; for, especially to an unpractised eye, it might be mistaken for the Yellow Hammer, and in some states of plumage for other Buntings. It has been taken in the neighbourhood of London. In 1837 there was a live specimen in an aviary of the Zoological Society in Regent's-park; and many Ortolans are sent alive to the London market from Prussia. There is, however, some consolation for the rarity of the Ortolan in England. It is approached in delicacy by our Wheatear, which is termed the *English Ortolan*. Hence it has been pursued as a delicate morsel throughout all its island haunts. Bewick captured it at sea, off the coast of Yorkshire, in May, 1822. Every spring and autumn it may be observed at Gibraltar, on its migration. Mr. Strickland saw it at Smyrna in April. North Africa is its winter residence. Colonel Sykes notes it in his catalogue of the birds of the Deccan.

Ortolans are solitary birds; they fly in pairs, rarely three together, and never in flocks. They are taken in traps from March or April to September, when they are often poor and thin; but if fed with plenty of millet-seed and other grain, they become sheer lumps of fat, and delicious morsels. They are fattened thus in large establishments in the south of Europe; Mr. Gould states this to be effected in Italy, and the south of France, in dark rooms; and the Prince of Musignano, having described the process, adds the relishing words, "Carne exquisita."

The fattening process in Italy is one of great refinement in the manner of feeding. It is the fat of the Ortolan which is so delicious; but it has a peculiar habit of feeding which is opposed to the rapid fattening, this is, it feeds only at the rising of the sun. Yet this peculiarity has not proved an insurmountable obstacle to the Italian gourmands. The Ortolans are placed in a dark chamber, perfectly dark, with only one aperture in the wall. The food is scattered over the floor of the chamber. At a certain hour in the morning the keeper of the birds places a lantern in the orifice of the wall; when the dim light thrown from the lantern on the floor of the apartment induces the Ortolans to believe that the sun is about to rise, and they greedily consume the food upon the floor. More food is now scattered over it, and the lantern is withdrawn.

The Ortolans, rather surprised at the shortness of the day, think it their duty to fall asleep, as night has spread her sable mantle round them. During sleep, little of the food being expended in the production of force, most of it goes to the formation of muscle and fat. After they have been allowed to repose for one or two hours, in order to complete the digestion of the food taken, their keeper again exhibits the lantern through the aperture. The "rising sun" a second time illumines the apartment, and the birds, awaking from their slumber, apply themselves voraciously to the food on the floor; after having discussed which, they are again enveloped in darkness. Thus the sun is made to shed its rising rays into the chamber floor four or five times every day, and as many nights following. The Ortolans thus treated become like little balls of fat in a few days. This not uninteresting process has been detailed by Dr. Lyon Playfair to the Royal Agricultural Society. It may, probably, be applied to purposes with less luxurious objects than fattening Ortolans.

Notwithstanding its delicacy, the Ortolan fattens very fast; and it is this lump of fatness that is its merit, and has sometimes caused it to be preferred to the Becafico. According to Buffon, the Greeks and Romans understood fattening the Ortolan upon millet. But a lively French commentator doubts this statement: he maintains that had the ancients known the Ortolan, they would have deified it, and built altars to it upon Mount Hymettus and the Saniculum; adding, did they not deify the horse of Caligula, which was certainly not worth an Ortolan? and Caligula himself, who was not worth so much as his horse? However, this dispute belongs to the "classics of the table."

The Ortolan is considered sufficiently fat when it is a handful, and is judged by feeling it, and not by appearance. It should not be killed with violence, like other birds; this might crush and bruise the delicate flesh, and spoil the *coup-d'œil*, to avoid which it is recommended to plunge the head of the Ortolan into a glass of brandy. The culinary instruction is as follows : having picked the bird of its feathers, singe it with the flame of paper or spirits of wine; cut off the beak and ends of the feet; do not draw it; put it into a paper case soaked in olive oil, and broil it over a slow fire of slack cinders, like that required for a pigeon *à la crapaudine*; in a few minutes the Ortolan will swim in its own fat, and will be cooked. Some gourmands wrap each bird in a vine-leaf.

A gourmand will take an Ortolan by the legs and craunch it in delicious mouthfuls, so as absolutely to lose none of it. More delicate feeders cut the bird into quarters, and lay aside the gizzard; the rest may be eaten, even to the bones, which are sufficiently tender for the most delicate mouth to masticate without inconvenience.

On the Continent, Ortolans are packed in tin boxes for exportation. They may be bought in London for half-a-crown a-piece. A few poulterers import Ortolans in considerable numbers, and some have acquired the art of fattening these birds.*

* The Ortolan figures in a curious anecdote of individual epicurism in the last century. A gentleman of Gloucestershire had one son, whom he sent abroad to make the grand tour of the Continent, where he paid more attention to the cookery of nations, and luxurious living, than anything else. Before his return his father died and left him a large fortune. He now looked over his note-book to discover where the most exquisite dishes were to be had, and the best cooks obtained. Every servant in his house was a cook; his butler, footman, coachman, and grooms—all were cooks. He had also three Italian cooks—one from Florence, another

Alexis Soyer put into the hundred guinea dish which he prepared for the royal table at the grand banquet at York, in 1850, five pounds worth of Ortolans, which were obtained from Belgium.

from Vienna, and another from Viterbo—for dressing one Florentine dish. He had a messenger constantly on the road between Britany and London to bring the eggs of a certain kind of plover found in the former country. This prodigal was known to eat a single dinner at the expense of 70*l.*, though there were but two dishes. In nine years he found himself getting poor, and this made him melancholy. When totally ruined, having spent 150,000*l.*, a friend one day gave him a guinea to keep him from starving, and he was found in a garret next day *broiling an Ortolan,* for which he had paid a portion of the alms.

TALK ABOUT TOUCANS.

THE Toucans, a family of climbing-birds of tropical America, appear to have been known in Europe by the length and great size of their bills, long before the birds themselves found their way to England. Belon, in 1555, described the bill of one of the family as half a foot long, large as a child's arm, pointed, and black at the tip, white elsewhere, notched on the edges, hollow within, and so finely delicate as to be transparent and thin as parchment; and its beauty caused it to be kept in the cabinets of the curious. For more than a century after Belon's work, the birds themselves had not been seen in England; for, in the *Museum Tradescantianum*, the standard collection of the time, and which, from the list of contributors, appears to have been the great receptacle for all curiosties, we read of an "Azacari (or Toucan) of Brazil; has his beak four inches long, almost two thick, like a Turk's sword" (A.D. 1656). From this description Tradescant knew the nature of the bird, if he had not seen it.

Mr. Swainson states, that the enormous bills give to these birds a most singular and uncouth appearance. Their feet are formed like those of the parrot, more for grasping than climbing; and as they live among trees, and proceed by hopping from branch to branch, their grasping feature is particularly adapted for such habits. They live retired in the deep forests, mostly in small companies. Their flight is strait and laborious, but not graceful; while their movements, as they glide rather than hop from branch to branch, are elegant.

Mr. Gould, in his grand Monograph of the Toucans, or *Ramphastidæ*, remarks, that it was only within a few years of the time of Linnæus that actual specimens of the Toucan had been received in Europe. The beaks, however, of these birds, regarded as curiosities, had occasionally found their way to our shores, and had occasioned some curious conjectures. The earliest shape resembled a Turkish scimitar.

The Toucans (a word derived from their Brazilian name, *Taca, Tucà*) received from Linnæus the title of *Ramphastos*, in allusion to the great volume of the beak (ραμφος—Ramphos), a family (*Ramphastidæ*). In some respects, indeed, they resemble the Hornbills in the development of the beak. The Toucans may be said to represent in America the Hornbills in India and Africa. Large as is the beak of the

Toucan compared with the size of the body, it is in reality very light. Its outer sheathing is somewhat elastic, very thin, smooth, and semi-transparent; and the interior consists of a maze of delicate cells, throughout which the olfactory nerves are multitudinously distributed. The nostrils are basal, the edges of each mandible are serrated, and the colouring of the whole beak is bright, rich, and often relieved by contrasted markings. But these tints begin to fade after death, and become ultimately dissipated. The eyes are surrounded by a considerable space of naked skin, often very richly tinted. The tongue is very long, slender, horizontally flattened, pointed, and, except at its base, horny; it is fringed or feathered along each side. The wings are short, concave, and comparatively feeble.

The tail is variable, equal and squared; it is remarkable for the facility with which it can be retroverted or turned up, so as to lie upon the back. This peculiarity results from a modification of structure in the caudal vertebræ, which enables the tail to turn with a jerk by the action of certain muscles, as if it were fixed on a hinge put into action by means of a spring. When the retroversion is accomplished, the muscles which caused it become passive, and offer no resistance to their antagonists, which restore the tail to its ordinary direction. When they sleep they puff out their plumage, they retrovert the tail over the back, draw the head between the shoulders; the bill begins to turn over the right shoulder, and becomes at last buried in the plumage of the back; at the same time the pinions of the wings droop, and conceal the feet. The bird now resembles an oval ball of puffed-up feathers, and is well protected against the cold.

Toucans utter, from time to time, harsh, clattering, and discordant cries. "Some," says Mr. Gould, "frequent the humid woods of the temperate regions, while others resort to comparatively colder districts, and dwell at an elevation of from six to ten thousand feet. Those inhabiting the lofty regions are generically different from those residing in the low lands, and are clothed in a more thick and sombre-coloured plumage. All the members of the Hill-Toucans are distinguished by their bills being strong, heavy, and hard, when compared with those of the true Toucans and Araçaris, all of which have their bills of a more delicate structure, and in several species so thin and elastic on the sides as to be compressible between the fingers." Their food in a state of nature consists of fruit, eggs, and nestling birds; to which, in domestication, are added small birds, mice, caterpillars, and raw flesh. They incubate in the hollows of gigantic trees.

Faber was told by Fryer, Alaysa, and other Spaniards who had lived long in America, and also by the Indians, that the Toucan even hews out holes in trees, in which to nidify; and Oviedo adds, that it is from this habit of chipping the trees that the bird is called by the Spaniards *Carpintero*, and by the Brazilians *Tacataca*, in imitation, apparently, of the sound it thus makes.

The larger feed upon bananas and other succulent plants; the smaller upon the smaller fruits and berries. Prince Maximilian de Wied states, that in Brazil he found only the remains of fruits in their stomachs, and adds, that they make sad havoc among plantations of fruit-trees. He was informed, however, that they steal and eat birds, but never himself saw them in the act. They abound in the vast forests, and are killed in great number in the cooler season in the year for the purposes of the table. In their manners the Toucans resemble the Crow tribe, and especially the Magpies: like them, they are very troublesome to the birds of prey, particularly to the Owls, which they surround, making a great noise, all the while jerking their tails upwards and downwards. Their feathers, especially from their yellow breasts, are used by the Indians for personal decoration.

Azara states that they attack even the solid nests of the white ants, when the clay of which their nests are formed becomes moistened with the rain; they break them up with their beaks, so as to obtain the young ants and their eggs; and during the breeding season the Toucan feeds upon nothing else; during the rest of the year he subsists upon fruit, insects, and the buds of trees.

Edwards, in his voyage up the Amazon, observes, that when a party of Toucans alight on a tree, one usually acts the part of a sentinel, uttering the loud cry of "Tucano," whence they derive their name; the others disperse over the branches in search of fruit. While feeding they keep up a hoarse chattering, and at intervals unite with the noisy sentry, and scream a concert that may be heard a mile. Having appeased their appetites, they seek the depths of a forest, and there quietly doze away the noon. In early morning a few of them may be seen sitting quietly upon the branches of some dead tree, apparently awaiting the coming sunlight before starting for their feeding-trees.

Some species of Toucans have been seen quarrelling with monkeys over a nest of eggs. Their carnivorous propensity has been strikingly shown in the specimens which have been kept in England. On the approach of any small bird the Toucan becomes highly excited, raises itself up, erects its feathers, and utters a hollow clattering sound, the irides of the eyes expand, and the Toucan is ready to dart on its prey. A Toucan, exhibited in St. Martin's-lane in 1824, seized and devoured a canary-bird. Next day Mr. Broderip tried him with a live goldfinch. The Toucan seized it with the beak, and the poor little victim uttered a short weak cry, for within a second it was dead, killed by the powerful compression of the mandibles. The Toucan now placed the dead bird firmly between its foot and the perch, stripped off the feathers with its bill, and then broke the bones of the wings and legs, by strongly wrenching them, the bird being still secured by the Toucan's foot. He then continued to work with great dexterity till he had reduced the goldfinch to a shapeless mass. This he devoured piece by piece with great gusto, not even leaving the legs or the beak of his prey: to each morsel he applied his tongue as he masticated it, chattering and shivering with delight. He never used his foot, but his bill, for conveying his food to his mouth by the sides of the bill.

Mr. Swainson remarks :—" The apparent disproportion of the bill is one of the innumerable instances of that beautiful adaptation of structure to use which the book of nature everywhere reveals. The food of these birds consists principally of the eggs and young of others, to discover which nature has given them the most exquisite powers of smell." Again, the nests in which the Toucan finds its food are often very deep and dark, and its bill, covered with branches of nerves, enables the bird to feel its way as accurately as the finest and most delicate finger could. From its feeding on eggs found in other birds' nests, it has been called the Egg-sucker. Probably there is no bird which secures her young offspring better from the monkeys, which are very noisome to the young of most birds. For when she perceives the approach of these enemies she so settles herself in her nest as to put her bill out at the hole, and give the monkeys such a welcome therewith that they presently break away, and are glad to escape.

Professor Owen, in his minute examination of the mandibles, remarks that the principle of the cylinder is introduced into the elaborate structure; the smallest of the supporting pillars of the mandibles are seen to be hollow or tubular when examined with the microscope.

Light and almost diaphonous as is the bill of the Toucan, its strength and the power of the muscles, which act upon the mandibles, are evident in the wrenching and masticatory processes. When taking fruit, the Toucan generally holds it for a short time at the extremity of his bill, applying to it, with apparent delight, the pointed tip of the slender tongue: the bird then throws it, with a sudden upward jerk, to the throat, where it is caught and instantly swallowed.

Mr. Gould divides the Toucans into six genera. 1. The true Toucans, with large and gaily-coloured bills, plumage black. 2. The Araçaris, with smaller beaks, plumage green, yellow, and red. 3. The Banded Aracauris, an Amazonian genus, proposed by Prince C. L. Bonaparte. 4. Toucanets, small, with crescent of yellow on the back, and brilliant orange and yellow ear-coverts. 5. Hill Toucans of the Andes. 6. Groove-bills, grass-green plumage.

A very fine true Toucan, figured by Mr. Gould, is remarkable for the splendour and size of the bill, of a fine orange-red, with a large black patch on each side. Powder-flasks are made of large and finely-coloured bills. The naked skin round the eye is bright orange. The chest is white, with a tinge of sulphur below, and a slight scarlet margin.

Upper tail-coverts, white; under tail-coverts, scarlet; the rest of the plumage, black. Several specimens of this beautiful bird lived both in the menagerie of the late Earl of Derby, at Knowsley, and in the gardens of the Zoological Society. It is a native of Cayenne, Paraguay, &c.

Toucans in their manners are gentle and confident, exhibiting no alarm at strangers, and are as playful as magpies or jackdaws; travellers assure us that they may be taught tricks and feats like parrots; and although they cannot imitate the human voice, they show considerable intelligence. One of the Toucanets is named from Mr. Gould, the plates in whose *monograph*, from their size, beauty, and accuracy, have all the air of portraits.

ECCENTRICITIES OF PENGUINS.

THIS group of amphibious birds, though powerless in the wing as an organ of flight, are assisted by it as a species of fin in their rapid divings and evolutions under water, and even as a kind of anterior of extremity when progressing on the land. Their lot has been wisely cast on those desolate southern islands and shores where man rarely intrudes, and in many instances where a churlish climate or a barren soil offers no temptations to him to invade their territory.

Le Vaillant, when on Dassen Island, found that the smaller crevices of the rocks served as places of retreat for Penguins, which swarmed there. "This bird," says Le Vaillant, which is about two feet in length, "does not carry its body in the same manner as others: it stands perpendicularly on its two feet, which gives it an air of gravity, so much the more ridiculous as its wings, which have no feathers, hang carelessly down on each side; it never uses them but in swimming. As we advanced towards the middle of the island we met innumerable troops of them. Standing firm and erect on their legs, these animals never deranged themselves in the least to let us pass; they more particularly surrounded the mausoleum, and seemed as if determined to prevent us from approaching it. All the environs were entirely beset with them. Nature had done more for the plain tomb of the poor Danish captain than what proceeds from the imaginations of poets or the chisels of our artists. The hideous owl, however well sculptured in our churches, has not half so dead and melancholy an air as the Penguin. The mournful cries of this animal, mixed with those of the sea-calf, impressed on my mind a kind of gloom which much disposed me to tender sensations of sadness. My eyes were sometime fixed on the last abode of the unfortunate traveller, and I gave his manes the tribute of a sigh."

Sir John Narborough says of the Patagonian Penguins that their erect attitude and bluish-black backs, contrasted with their white bellies, might cause them to be taken at a distance for young children with white pinafores on. A line of them is engraved in Webster's "Voyage of the Chanticleer," and reminds us of one of the woodcuts in Hood's "Comic Annual."

The "towns, camps, and rookeries," as they have been called, of Penguins have been often described. At the Falkland Islands are assemblies of Penguins, which give a dreary desolation to the place, in the utter absence of the human race. In some of the towns voyagers describe a general stillness, and when the intruders walked among the feathered population to provide themselves with eggs, they were regarded with side-long glances, but they seemed to carry no terror with them. In many places the shores are covered with these birds, and three hundred have been taken within an hour; for they generally make no effort to escape, but stand quietly by whilst their companions are knocked down with sticks, till it comes to their turn.

The rookeries are described as designed with the utmost order and regularity, though they are the resort of several different species. A regular camp, often covering three or four acres, is laid out and levelled, and the ground disposed in squares for the nests, as accurately as if a surveyor had been employed. Their marchings and countermarchings are said to remind the observer of the manoeuvres of soldiers on parade. In the midst of this apparent order there appears to be not very good government, for the stronger species steal the eggs of the weaker if they are left unguarded; and the King Penguin is the greatest thief of all. Three species are found in the Falkland Islands. Two, the *Kings* and the *Macaroni*, deposit their eggs in these rookeries. The *Jackass*, which is the third, obtained its English name from its brayings at night. It makes its nests in burrows on downs and sandy plains; and Forster describes the ground as everywhere so much bored, that a person, in walking, often sinks up to the knees; and if the Penguin chance to be in her hole, she revenges herself on the passenger by fastening on his legs, which she bites very hard.

But these rookeries are insignificant when compared with a settlement of King Penguins, which Mr. G. Bennett saw at the north end of Macquarrie Island, in the South Pacific Ocean—a colony of these birds, which covered some thirty or forty acres. Here, during the whole of the day and night, 30,000 or 40,000 Penguins are continually landing, and an equal number going to sea. They are ranged, when on shore, in as regular ranks as a regiment of soldiers, and are classed, the young birds in one situation, the moulting birds in another, the sitting hens in a third, the clean birds in a fourth, &c.; and so strictly do birds in a similar condition congregate, that, should a moulting bird intrude itself among those which are clean, it is immediately ejected from them. The females, if approached during incubation, move away, carrying their eggs with them. At this time the male bird goes to sea, and collects food for the female, which becomes very fat.

Captain Fitzroy describes, at Noir Island, multitudes of Penguins swarming among the bushes and tussac-grass near the shore, for moulting and rearing their young. They were very valiant in self-defence, and ran open-mouthed by dozens at any one who invaded their territory. The manner of feeding their young is amusing. The old bird gets on a little eminence and makes a loud noise, between quacking and braying, holding its head up as if haranguing the Penguinnery, the young one standing close to it, but a little lower. The old bird then puts down its head, and opens its mouth widely, into which the young one thrusts its head, and then appears to suck from the throat of its mother; after which the clatter is repeated, and the young one is again fed: this continues for about ten minutes.

Mr. Darwin, having placed himself between a Penguin, on the Falkland Islands, and the water, was much amused by watching its habits. "It was a brown bird," says Mr. Darwin, "and, till reaching the sea, it regularly fought and drove me backwards. Nothing less than heavy blows would have stopped him: every inch gained, he firmly kept standing close before me, erect and determined. When thus opposed, he continually rolled his head from side to side in a very odd manner. While at sea, and undisturbed, this bird's note is very deep and solemn, and is often heard in the night time. In diving, its little plumeless wings are used as fins, but on the land as front legs. When crawling (it may be said on four legs) through the tussacks, or on the side of a grassy cliff, it moved so very quickly that it might readily have been mistaken for a quadruped. When at sea, and fishing, it comes to the surface for the purpose of breathing with such a spring, and dives again so instantaneously, that I defy any one, at first sight, to be sure that it is not a fish leaping for sport."

Bougainville endeavoured to bring home a Penguin

alive. It became so tame that it followed the person who fed it; it ate bread, flesh, or fish; but it fell away and died. The four-footed Duck of Gesner might have owed its origin to an ill-preserved Penguin. The notion of its being four-footed might have been fortified by some voyager who had seen the bird making progress as Mr. Darwin has above described.

Mr. Webster describes the feathers of Penguins as very different from those of other birds, being short, very rigid, and the roots deeply imbedded in fat. They are, in general, flat, and bent backwards, those on the breast being of a satin or silky white, and those on the flippers so short and small as to approach the nature of scales, overlaying each other very closely. The skins are loaded with fat. Their feet are not regularly webbed, but present a broad, fleshy surface, more adapted for walking than swimming. Mr. Webster saw great numbers of Penguins on Staten Island. They are the only genus of the feathered race that are there, and live in the water, like seals. He saw them at the distance of 200 miles from the land, swimming with the rapidity of the dolphin, the swiftest of fishes. When they come up to the surface for fresh breath, they make a croaking noise, dip their beaks frequently in the water, and play and dive about near the surface, like the bonita. Penguins have great powers of abstinence, and are able to live four or five months without food. Stones have been occasionally found in their stomachs, but they generally live on shrimps and crustacea, gorging themselves sometimes to excess. The sensations of these curious birds do not seem to be very acute. Sparrman stumbled over a sleeping one, and kicked it some yards, without disturbing its rest; and Forster left a number of Penguins apparently lifeless, while he went in pursuit of others, but they afterwards got up and marched off with their usual gravity.

The bird is named from the Welsh word, *Pengwyn*, White head (*pen*, head; *gwyn*, white), and is thought to have been given to the bird by some Welsh sailors, on seeing its white breast. Davis, who discovered, in 1585, the straits which are named after him, was of Welsh parents. Might he not have given the name *Pengwyn* to the bird? Swainson considers the Penguins, on the whole, as the most singular of all aquatic birds; and he states that they clearly point out that nature is about to pass from the birds to the fishes. Others consider Penguins more satisfactorily to represent some of the aquatic reptiles, especially the marine *testudinata*.

PELICANS AND CORMORANTS.

PELICANS are described as a large, voracious, and wandering tribe of birds, living for the most part on the ocean, and seldom approaching land but at the season of incubation. They fly with ease, and even with swiftness. Their bill is long, and armed at the end with an abrupt hook; the width of the gape is excessive; the face is generally bare of feathers, and the skin of the throat sometimes so extensible as to hang down like a bag; it will occasionally contain ten quarts. "By this curious organization," observes Swainson, "the Pelicans are able to swallow fish of a very large size; and the whole family may be termed *oceanic vultures.*"

The neighbourhood of rivers, lakes, and the sea-coast, is the haunt of the Pelican, and they are rarely seen more than twenty leagues from the land. Le Vaillant, upon visiting Dassen Island, at the entrance of Saldanha Bay, beheld, as he says, after wading through the surf, and clambering up the rocks, such a spectacle as never, perhaps, appeared to the eye of mortal. "All of a sudden there arose from the whole surface of the island an impenetrable cloud, which formed, at the distance of forty feet above our heads, an immense canopy, or, rather, a sky, composed of birds of every species and of all colours—cormorants, sea-gulls, sand-swallows, and, I believe, the whole winged tribe of this part of Africa, were here assembled." The same traveller found on the Klein-Brak river, whilst waiting for the ebb-tide, thousands of Pelicans and Flamingoes, the deep rose-colour of the one strongly contrasting with the white of the other.

Mr. Gould says the bird is remarkable for longevity and the long period requisite for the completion of its plumage. The first year's dress is wholly brown, then fine white. The rosy tints are only acquired as the bird advances in age, and five years are required before the Pelican becomes fully mature. The expanse of wings is from twelve to thirteen feet. Although the bird perches on trees, it prefers rocky shores. It is found in the Oriental countries of Europe; and is common on the rivers and lakes of Hungary and Russia, and on the Danube. That the species exists in Asia there is no doubt. Belon, who refers to Leviticus xi. 18, where the bird is noted as unclean, says that it is frequent on the lakes of Egypt and Judæa. When he was passing the plain of Roma, which is only half a day's journey from Jerusalem, he saw them flying in pairs, like swans, as well as in a large flock. Hasselquist saw the

Pelican at Damietta, in Egypt. "In flying, they form an acute angle, like the common wild geese when they migrate. They appear in some of the Egyptian drawings."—(*Rossellini.*)

Von Siebold saw the Pelican in Japan. "Pelicans," says Dr. Richardson, "are numerous in the interior of the fur countries, but they seldom come within two hundred miles of Hudson's Bay. They deposit their eggs usually on small rocky islands, on the brink of cascades, where they can scarcely be approached; but they are otherwise by no means shy birds. They haunt eddies under waterfalls, and devour great quantities of carp and other fish. When gorged with food they doze on the water, and may be easily captured, as they have great difficulty in taking wing at such times, particularly if their pouches be loaded with fish."

The bird builds on rocky and desert shores : hence we read of "the Pelican of the wilderness," alluded to in these beautiful lines :—

> " Like the Pelicans
> On that lone island where they built their nests,
> Nourish'd their young, and then lay down to die."

The bird lives on fish, which it darts upon from a considerable height. James Montgomery thus describes this mode of taking their prey :—

> " Eager for food, their searching eyes they fix'd
>   On Ocean's unroll'd volume, from a height
>   That brought immensity within their scope ;
>   Yet with such power of vision look'd they down,
>   As though they watch'd the shell-fish slowly gliding

O'er sunken rocks, or climbing trees of coral.
On indefatigable wing upheld,
Breath, pulse, existence, seem'd suspended in them ;
They were as pictures painted on the sky ;
Till suddenly, aslant, away they shot,
Like meteors chang'd from stars in gleams of lightning,
And struck upon the deep ; where, in wild play,
Their quarry flounder'd, unsuspecting harm.
With terrible voracity they plunged
Their heads among the affrighted shoals, and beat
A tempest on the surges with their wings,
Till flashing clouds of foam and spray conceal'd them.
Nimbly they seized and secreted their prey,
Alive and wriggling, in th' elastic net
Which Nature hung beneath their grasping beaks ;
Till, swoll'n with captures, th' unwieldy burthen
Clogg'd their slow flight, as heavily to land
These mighty hunters of the deep return'd.
There on the cragged cliffs they perched at ease,
Gorging their hapless victims one by one ;
Then, full and weary, side by side they slept,
Till evening roused them to the chase again."

Pelican Island.

Great numbers of Pelicans are killed for their pouches, which are converted by the native Americans into purses, &c. When carefully prepared, the membrane is as soft as silk, and sometimes embroidered by Spanish ladies for work-bags, &c. It is used in Egypt by the sailors, whilst attached to the two under chaps, for holding or baling water.

With the Pelican has been associated an old popular error, which has not long disappeared from books of information: it is that of the Pelican feeding her young with her blood. In reference to the actual economy of the Pelican, we find that, in feeding the nestlings—and the male is said to supply the wants of the female, when sitting, in the same manner—the under mandible is pressed against the neck and breast, to assist the bird in disgorging the contents of the capacious pouch; and during this action the red nail of the upper mandible would appear to come in contact with the breast, thus laying the foundation, in all probability, for the fable that the Pelican nourishes her young with her blood, and for the attitude in which the imagination of painters has placed the bird in books of emblems, &c., with the blood spirting from the wounds made by the terminating nail of the upper mandible into the gaping mouths of her offspring.

Sir Thomas Browne, in his "Vulgar Errors," says:—" In every place we meet with the picture of the Pelican opening her breast with her bill, and feeding her young ones with the blood distilling from her. Thus it is set forth, not only in common signs, but in the crest and scutcheon of many noble families; hath been asserted by many holy writers, and was an hieroglyphic of piety and pity among the Egyptians; on which consideration they spared them at their tables."

Sir Thomas refers this popular error to an exaggerated description of the Pelican's fondness for her young, and is inclined to accept it as an emblem "in coat-armour," though with great doubt.

In "A Choice of Emblems and other Devices," by Geoffrey Whitney, are these lines:—

> "The Pelican, for to revive her younge,
>   Doth pierce her breste, and geve them of her blood.
> Then searche your breste, and as you have with tonge,
>   With penne procede to do your countrie good:
> Your zeal is great, your learning is profounde;
> Then help our wantes with that you do abound."

In George Wither's "Emblems," 1634, we find—

> "Our Pelican, by bleeding thus,
>   Fulfill'd the law, and cured us."

Shakspeare, in "Hamlet," thus alludes to the popular notion:—

> "To his good friends thus wide I'll ope my arms;
> And like the kind, life-rendering Pelican,
> Repast them with my blood."

In a holier light, this symbol signifies the Saviour giving Himself up for the redemption of mankind. In Lord Lindsay's "Christian Art," vols. i., xx., xxi., we find in the text, "God the Son (is symbolized) by a Pelican—'I am like a Pelican of the wilderness.' (Psalm cii. 6.)" To which is added the following note:—"The mediæval interpretation of this symbol is given by Sir David Lindsay, of the Mount, Lion King, nephew of the poet, in his MS. 'Collectanea,' preserved in the Advocates' Library, Edinburgh.

Sir Thomas Browne hints at the probability of the Pelican occasionally nibbling or biting itself on the itching part of its breast, upon fulness or acrimony of blood, so as to tinge the feathers in that part. Such an instance is recorded by Mr. G. Bennett of a

Pelican living at Dulwich, which wounded itself just above the breast; but no such act has been observed among the Pelicans kept in the menagerie of the Zoological Society or elsewhere; and the instance just recorded was probably caused by local irritation.

Of the same genus as the *Pelican* is the *Cormorant*, an inhabitant of Europe generally and of America. It swims very deep in the water; even in the sea very little more than the neck and head are visible above the surface. It is a most expert diver, pursuing the fish which forms its food with great activity under water; it is said to be very fond of eels. It perches on trees, where it occasionally builds its nests, but it mostly selects rocky shores and islands. Upon the Fern Islands its nest is composed of a mass of sea-weed, frequently heaped up to the height of two feet. The species is easily domesticated; and its docility is shown by the use often made of Cormorants in fishing. Willughby, quoting Faber, says:—" They are wont in England to train up Cormorants to fishing. When they carry them out of the room where they are kept they take off their hoods, and having tied a leather thong round the lower part of their necks, that they may not swallow down the fish they catch, they throw them into the river. They presently dive under water, and there for a long time, with wonderful swiftness, pursue the fish, and when they have caught them they arise presently to the top of the water, and pressing the fish tightly with the bills, they swallow them, till each bird hath after this manner devoured five or six fishes. Then their keepers call them to the fish, to which they readily fly, and little by little, one after another, vomit up all the fish, a little bruised with the nip they gave them with their bills." When they have done fishing they loosen the string from the birds' necks, and for their reward they throw them part of the prey they have caught, to each, perchance, one or two fishes, which they catch most dexterously in their mouths as they are falling in the air. Pennant quotes Whitelock, who said that he had a cast of them, manned like hawks, and which would come to hand. He took much pleasure in them, and relates that the best he had was one presented him by Mr. Wood, master of the corvorants (as the older name was) to Charles I. Pennant adds, it is well known that the Chinese make great use of a congenerous sort in fishing, and that not for amusement but profit.

Sir George Staunton, in his account of his Embassy to China, describes the place where the *Leu-tze*, or famed fishing-bird of China, is bred and instructed in the art and practice of supplying his owner with fish in great abundance. The bird, a Cormorant, is figured in Sir George's work, with two Chinese fishermen carrying their light boat, around the gunnel of which their Cormorants are perched by a pole resting on their shoulders between them. On a large lake are thousands of small boats and rafts built entirely for this species of fishery. On each boat or raft are ten or a dozen birds, which, on a signal from the owner, plunge into the water; and it is astonishing to see the enormous size of fish with which they return grasped between their bills. They appeared to be so well trained that it did not require either ring or cord about their throats to prevent them from swallowing any portion of their prey except what the master was pleased to return to them for encouragement and food. The boat used by these fishermen is remarkably light, and is often carried to the lake, together with the fishing-birds, by the men who are there to be supported by it.

Belon gives an amusing account of the chase of this bird during calms, especially in the neighbourhood of Venice: the hunt is carried on in very light boats, each of which being rowed by five or six men, darts along the sea like the bolt from an arbalest, till the poor Cormorant, who is shot at with bows as soon as he puts his head above water, and cannot take flight after diving to suffocation, is taken quite tired out by his pursuers.

Cormorant fishing has occasionally been reintroduced upon our rivers. In 1848 there were brought from Holland four tame Cormorants, which had been trained to the Chinese mode of fishing. Upon one occasion they fished three miles on a river, and caught a pannier-full of trout and eels. A ring si placed round their necks to prevent them from swallowing large fish, but which leaves them at liberty to gulp down anything not exceeding the size of a gudgeon. The birds on these occasions are put into such parts of the river as are known to be favourite haunts of fish; and their activity under water in pursuit of fish can be compared to nothing so appropriate as a swallow darting after a fly.

Blumenbach tells us the Cormorant occasionally increases in a few years to many thousands on coasts where it was previously unknown. It varies much both in size and colour. The late Joshua Brookes, the surgeon, possessed a Cormorant, which he presented to the Zoological Society.

The Cormorant has a small sabre-shaped bone at the back of its vertex; which bone may serve as a lever in throwing back the head, when the animal tosses the fishes into the air and catches them in its open mouth. The same motion is, however, performed by some piscivorous birds, which are not provided with this particular bone.

Aubrey, in his "Natural History of Wilts," quotes the following weather presage from May's "Virgil's Georgics":—

> "The seas are ill to sailors evermore
> When Cormorants fly crying to the shore."

TALKING BIRDS, ETC.

ERTAIN birds are known to utter strange sounds, the origin of which has much puzzled the ornithologists. The Brown Owl which hoots, is hence called the Screech Owl: a musical friend of Gilbert White tried all the Owls that were his near neighbours with a pitch-pipe set at a concert pitch, and found they all hooted in B flat; and he subsequently found that neither Owls nor Cuckoos keep to one note. The Whidah Bird, one of the most costly of cage-birds, rattles its tail-feathers with a noise somewhat resembling that made by the rattle-snake. The Chinese Starling, in China called *Longuoy*, in captivity is very teachable, imitating words, and even whistling tunes: we all remember Sterne's Starling. The Piping Crow, to be seen in troops in the Blue Mountains, is named from its ready mimicry of other birds: its imitation of the chucking and cackling of a hen and the crowing of a cock, as well as its whistling of tunes, are described as very perfect: its native

note is said to be a loud whistle. The Blue Jay turns his imitative faculty to treacherous account : he so closely imitates the St. Domingo Falcon as to deceive even those acquainted with both birds; and the Falcon no sooner appears in their neighbourhood than the jays swarm around him and insult him with their imitative cries; for which they frequently fall victims to his appetite. The Bullfinch, according to Blumenbach, learns to whistle tunes, to sing in parts, and even to pronounce words. The note of the Crowned Crane has been compared by Buffon to the hoarseness of a trumpet; it also clucks like a hen. Mr. Wallace, in his "Travels on the Amazon," saw a bird about the size and colour of the Raven, which uttered a loud, hoarse cry, like some deep musical instrument, whence its Indian name, *Ueramioube*, Trumpet Bird : it inhabits the flooded islands of the Rio Negro and the Solimoes, never appearing on the mainland.* The only sound produced by Storks

---

* The popular name of this bird is the *Umbrella Bird*. On its head it bears a crest, different from that of any other bird. It is formed of feathers more than two inches long, very thickly set, and with hairy plumes curving over at the end. These can be laid back so as to be hardly visible, or can be erected and spread out on every side, forming a dome completely covering the head, and even reaching beyond the point of the beak ; the individual feathers then stand out something like the down-bearing seeds of the dandelion. Besides this, there is another ornamental appendage on the breast, formed by a fleshy tubercle, as thick as a quill and an inch and a-half long, which hangs down from the neck, and

is by snapping their bills. The Night Heron is called the Qua Bird, from its note *Qua*.

The Bittern, the English provincial names of which are the Mire-drum, Bull of the Bog, &c., is so called for the bellowing or drumming noise or booming for which the bird is so famous. This deep note of the "hollow-sounding Bittern" is exerted on the ground at the breeding season, about February or March. As the day declines he leaves his haunt, and, rising spirally, soars to a great height in the twilight. Willughby says that it performs this last-mentioned feat in the autumn, "making a singular kind of noise, nothing like to lowing." Bewick says that it soars as above described when it changes its haunts. Ordinarily it flies heavily, like the Heron, uttering from time to time a resounding cry, not bellowing; and then Willughby, who well describes the bellowing noise of the breeding season, supposes it to be the Night Raven, at whose "deadly voice" the superstitious wayfarer of the night turned pale and trembled. "This, without doubt," writes Willughby, "is that bird our common people call the Night Raven, and have such a dread of, imagining its cry portends no less than their death or the death of some of their near relations; for it flies in the night, answers their description of being like a flagging collar, and hath such a kind of hoop-

is thickly covered with glossy feathers, forming a large pendent plume or tassel. This, also, the bird can either press to its breast, so as to be scarcely visible, or can swell out so as almost to conceal the forepart of its body.

ing cry as they talk of." Others, with some reason, consider the Qua Bird already mentioned (which utters a loud and most disagreeable noise when on the wing, conveying the idea of the agonies of a person attempting to vomit) to be the true Night Raven. The Bittern was well known to the ancients, and Aristotle mentions the fable of its origin from staves metamorphosed into birds. The long claw of the hind toe is much prized as a toothpick, and in the olden times it was thought to have the property of preserving the teeth.

The Greater-billed Butcher Bird, from New Holland, has extraordinary powers of voice: it is trained for catching small birds, and it is said to imitate the notes of some other birds by way of decoying them to their destruction.

The mere imitative sounds of Parrots are of little interest compared with the instances of instinct, apparently allied to reason, which are related of individuals. Of this tribe the distinguishing characteristics are a hooked bill, the upper mandible of which is moveable as well as the lower, and not in one piece with the skull, as in most other birds, but joined to the head by a strong membrane, with which the bird lifts it or lets it fall at pleasure. The bill is also round on the outside and hollow within, and has, in some degree, the capacity of a mouth, allowing the tongue, which is thick and fleshy, to play freely; while the sound, striking against the circular border of the lower mandible, reflects it like a palate: hence the animal does not utter a whistling sound, but a full articulation. The tongue, which modulates all sounds, is proportionally larger than in man.

The Wild Swan has a very loud call, and utters a melancholy cry when one of the flock is killed; hence it was said by the poets to sing its own dying dirge. Such was the popular belief in olden times; and, looking to the anatomical characteristics of the species, it was, in some degree, supported by the more inflated windpipe of the wild when compared with that of the tame species. The *Song of the Swan* is, however, irreconcileable with sober belief, the only noise of the Wild Swan of our times being unmelodious, and an unpleasing monotony.

The Laughing Goose is named from its note having some resemblance to the laugh of man; and not, as Wilson supposes, from the grinning appearance of its mandibles. The Indians imitate its cry by moving the hand quickly against the lips, whilst they repeat the syllable *wah*.

The Cuckoo may be said to have done much for musical science, because from that bird has been derived the *minor scale*, the origin of which has puzzled so many; the Cuckoo's couplet being the *minor third* sung downwards.

The Germans are the finest appreciators of the Nightingale; and it is a fact, that when the Prussian authorities, under pecuniary pressure, were about to cut down certain trees near Cologne, which were frequented by Nightingales, the alarmed citizens purchased the trees in order to save the birds and keep their music. Yet one would think the music hardly worth having, if it really sounded as it looks upon paper, transcribed thus by Bechstein, from whom it is quoted by Broderip:—

Zozozozozozozozozozozozo zirrhading
Hezezezezezezezezezezezezezeze cowar ho dze hoi
Higaigaigaigaigaigaigaigaigaigaigai, guaiagai coricor dzio dzio pi.*

M. Wichterich, of Bonn, remarks:—"It is a vulgar error to suppose that the song of the Nightingale is melancholy, and that it only sings by night. There are two varieties of the Nightingale; one which sings both in the night and the day, and one which sings in the day only."

In the year 1858, Mr. Leigh Sotheby, in a letter to Dr. Gray, of the British Museum, described a marvellous little specimen of the feathered tribe—a Talking Canary. Its parents had previously and successfully reared many young ones, but three years before they hatched only *one* out of four eggs, the which they immediately neglected, by commencing the rebuilding of a nest on the top of it. Upon this discovery, the unfledged and forsaken bird, all but dead, was taken away and placed in flannel by the fire, when, after much attention, it was restored, and then brought up by hand. Thus treated, and away from all other birds, it became familiarised only with those who fed it; consequently, its first singing notes were of a character totally different to those usual with the Canary.

* "Athenæum," No. 1467.

Constantly being talked to, the bird, when about three months old, astonished its mistress by repeating the endearing terms used in talking to it, such as "Kissie, kissie," with its significant sounds. This went on, and from time to time the little bird repeated other words; and then, for hours together, except during the moulting season, it astonished by *ringing the changes*, according to its own fancy, and as plainly as any human voice could articulate them, on the several words, "Dear sweet Titchie" (its name), "kiss Minnie," "Kiss me, then, dear Minnie," "Sweet pretty little Titchie," "Kissie, kissie, kissie," "Dear Titchie," "Titchie wee, gee, gee, gee, Titchie, Titchie."

The usual singing-notes of the bird were more of the character of the Nightingale, mingled occasionally with the sound of the dog-whistle used about the house. It is hardly necessary to add, that the bird was by nature remarkably tame.

In 1839, a Canary-bird, capable of distinct articulation, was exhibited in Regent-street. The following were some of its sentences:—"Sweet pretty dear," "Sweet pretty dear Dicky," "Mary," "Sweet pretty little Dicky dear;" and often in the course of the day, "Sweet pretty Queen." The bird also imitated the jarring of a wire, the ringing of a bell; it was three years old, and was reared by a lady who never allowed it to be in the company of other birds. This Canary died in October, 1839; it was, it is believed, the only other talking instance publicly known.

We read of some experiments made in the rearing

of birds at Kendal by a bird-fancier, the result of which was, that upwards of 20 birds — Canaries, Greenfinches, Linnets, Chaffinches, Titlarks, and Whitethroats—were reared in one cage by a pair of Canaries. The experiments were continued until the extraordinary number of thirty-eight birds had been brought up within two months by the Canaries. It may be worth while to enumerate them.

In the month of June the Canaries—the male green, and the female piebald—were caged for the purpose of breeding. The female laid five eggs, and while she was sitting a Greenfinch egg was introduced into the nest. All of these were hatched, and the day after incubation was completed five Grey Linnets, also newly hatched, were put into the cage, in their own nest. Next day a newly-hatched nest of four Chaffinches was also introduced; and afterwards five different nests, consisting of six Titlarks, six Whitethroats, three Skylarks, three Winchars, and three Blackcaps. While rearing the last of these nests, the female Canary again laid and hatched four eggs, thus making thirty-eight young birds brought up by the pair of Canaries. It will be noticed that most of these birds are soft-billed, whose natural food is small insects; but they took quite kindly to the seeds upon which they were fed by their step-parents. The pair of Canaries fed at one time twenty-one young birds, and never had less than sixteen making demands upon their care; and while the female was hatching her second nest she continued to feed the birds that occupied the other nest.

Of the origin of the *neighing sound* which accompanies the single Snipe's play-flight during pairing-time, opinions are various. Bechstein thought it was produced by means of the beak; Naumann and others, again, that it originated in powerful strokes of the wing. Pratt, in Hanover, observing that the bird makes heard its well-known song or cry, which he expresses with the words, "gick jack, gick jack!" at the same time with the *neighing sound*, it seemed to be settled that the latter is not produced through the throat. In the meantime, M. Meves, of Stockholm, remarked with surprise, that the humming sound could never be observed whilst the bird was flying upwards, at which time the tail is closed; but only when it was casting itself downwards in a slanting direction, with the tail strongly spread out.

M. Meves has written for the Zoological Society a paper upon the origin of this sound, which all the field-naturalists and sportsmen of England and other countries had, for the previous century, been trying to make out, but had failed to discover. Of this paper the following is an abstract:—

The peculiar form of the tail-feathers in some foreign species nearly allied to our Snipe encouraged the notion that the tail conduced to the production of the sound. M. Meves found the tail-feathers of our common Snipe, in the first feather especially, very peculiarly constructed; the shaft uncommonly stiff and sabre-shaped; the rays of the web strongly bound together and very long, the longest reaching nearly three-fourths of the whole length of the web, these rays lying along or spanning from end to end of the curve of the shaft, *like the strings of a musical instrument.* If you blow from the outer side upon the broad web, it comes into vibration, and a sound is heard, which, though fainter, resembles very closely the well-known *neighing*.

But to convince yourself fully that it is the first feather which produces the peculiar sound, it is only necessary carefully to pluck out such an one, to fasten its shaft with fine thread to a piece of steel wire a tenth of an inch in diameter, and a foot long, and then to fix this at the end of a four-foot stick. If now you draw the feather, with this outer side forward, sharply through the air, at the same time making some short movements or shakings of the arm, so as to represent the shivering motion of the wings during flight, you produce the neighing sound with the most astonishing exactness.

If you wish to hear the humming of both feathers at once, as must be the case from the flying bird, this also can be managed by a simple contrivance. Take a small stick, and fasten at the side of the smaller end a piece of burnt steel wire in the form of a fork; bind to each point a side tail-feather; bend the wire so that the feathers receive the same direction which they do in the spreading of the tail as the bird sinks itself in flight; and then, with this apparatus, draw the feathers through the air as before. Such a sound, but in another tone, is produced when we experiment with the tail-feathers of other kinds of Snipe.

Since in both sexes these feathers have the same form, it is clear that both can produce the same humming noise; but as the feathers of the hen are generally less than those of the cock-bird, the noise made by them is not so deep as in the other case.

Besides the significance which these tail-feathers have as a kind of musical instrument, their form may give a weighty character in the determination of a species standing very near one another, which have been looked upon as varieties.

This interesting discovery was first announced by M. Meves in an account of the birds observed by himself during a visit to the Island of Gottland, in the summer of the year 1856, which narrative was published at Stockholm in the following winter. In the succeeding summer, M. Meves showed his experiments to Mr. Wolley, whose services to Ornithology we have already noticed. The mysterious noise of the wilderness was reproduced in a little room in the middle of Stockholm: first, the deep bleat, now shown to proceed from the male Snipe, and then the fainter bleat of the female, both most strikingly true to nature, neither producible with any other feathers than the outer ones of the tail.

Mr. Wolley inquired of Mr. Meves how, issuing forth from the town on a summer ramble, he came to discover what had puzzled the wits and strained the eyes of so many observers. He freely explained how, in a number of "Naumannia," an accidental misprint of the word representing tail-feathers instead of wing-feathers,—a mistake which another author ridiculed—first led him to think on the subject. He subsequently examined in the Museum at Stockholm the tail-feathers of various species of Snipe, remarked their structure, and reasoned upon it. Then he blew upon them, and fixed them on levers that he might wave them with greater force through the air; and at the same time he made more careful observation than he had hitherto done in the living birds. In short, in him the obscure hint was thrown upon fruitful ground, whilst in a hundred other minds it had failed to come to light.

Dr. Walsh saw at Constantinople a Woodpecker, about the size of a Thrush, which was very active in devouring flies, and tapped woodwork with his bill with a noise *as loud as that of a hammer*, to disturb the insects concealed therein, so as to seize upon them when they appeared.

Among remarkable bird services should not be forgotten those of the Trochilos to the Crocodile. "When the Crocodile," says Herodotus, "feeds in the Nile, the inside of his mouth is always covered with *bdella* (a term which the translators have rendered by that of *leech*). All birds, *except one*, fly from the Crocodile; but this one bird, the *Trochilos*, on the contrary, flies towards him with the greatest eagerness, and renders him a very great service; for every time that the Crocodile comes to the land to sleep, and when he lies stretched out with his jaws open, the Trochilos enters and establishes himself in his mouth, and frees him from the bdella which he finds there. The Crocodile is grateful, and never does any harm to the little bird who performs for him this office."

This passage was long looked upon as a pleasant story, and nothing more; until M. Geoffroy St. Hilaire, during his long residence in Egypt, ascertained the story of Herodotus to be correct in substance, but inexact in details. It is perfectly true that a little bird does exist, which flies incessantly from place to place, searching everywhere, even in the Crocodile's mouth, for the insects which form the principal part of its nourishment. This bird is seen everywhere on the banks of the Nile, and M. Geoffroy has proved it to be of a species already described by Hasselquist, and very like the small winged Plover. If the Trochilos be in reality the little Plover, the bdella cannot be leeches, (which do not exist in the running waters of the Nile) but the small insects known as *gnats* in Europe. Myriads of these insects dance upon the Nile: they attack the Crocodile upon the inner surface of his palate, and sting the orifice of the glands, which are numerous in the Crocodile's mouth. Then the little Plover, who follows him everywhere, delivers him from these troublesome enemies; and that without any danger to himself, for the Crocodile is always careful, when he is going to shut his mouth, to make some motion which warns the little bird to fly away. At St. Domingo there is a Crocodile which very nearly resembles that of Egypt. This Crocodile is attacked by gnats, from which he would have no means of delivering himself (his tongue, like that of the Crocodile, being fixed) if a bird of a particular species did not give him the same assistance that the Crocodile of the Nile receives from the little Plover. These facts explain the passage in Herodotus, and demonstrate that the animal, there called bdella, is not a leech, but a flying insect similar to our gnat.

Exemplifications of instinct, intelligence, and reason in Birds are by no means rare, but this distinction must be made: instinctive actions are dependent on the nerves, intelligence on the brain; but that which constitutes peculiar qualities of the mind in man has no material organ. The Rev. Mr.

Statham has referred to the theory of the facial angle as indicative of the amount of sagacity observable in the animal race, but has expressed his opinion that the theory is utterly at fault in the case of Birds; many of these having a very acute facial angle being considerably more intelligent than others having scarcely any facial angle at all. Size also seems to present another anomaly between the two races of Beasts and Birds; for while the Elephant and the Horse are among the most distinguished of quadrupeds for sagacity and instinct, the larger Birds seem scarcely comparable to the smaller ones in the possession of these attributes. The writer instances this by comparing the Ostrich and the Goose with the Wren, the Robin, the Canary, the Pigeon, and the Crow; and amusingly alludes to the holding of parliaments or convocations of birds of the last species, while the Ostrich is characterised in Scripture as the type of folly.

The author then refers to the poisoning of two young Blackbirds by the parent birds, when they found that they could neither liberate them nor permanently share their captivity. The two fledglings had been taken from a Blackbirds' nest in Surrey-square, and had been placed in a room looking over a garden, in a wicker cage. For some time the old birds attended to their wants, visited them regularly, and fed them with appropriate food; but, at last, getting wearied of the task, or despairing of effecting their liberation, they appear to have poisoned them. They were both found suddenly dead one morning, shortly after having been seen in good health; and on opening their bodies a small leaf, supposed to be that of *Solanum Nigrum*, was found in the stomach of each. The old birds immediately deserted the spot, as though aware of the nefarious deed befitting their name.

As an exemplification of instinct Dr. Horner states that Rooks built on the Infirmary trees at Hull, but never over the street. One year, however, a young couple ventured to build here: for eight mornings in succession the old Rooks proceeded to destroy the nest, when at last the young ones chose a more fitting place.

Mr. A. Strickland, having referred to the tendency of birds to build their nests of materials of a colour resembling that around such nests, relates an instance in which the Fly-catcher built in a red brick wall, and used for the nest mahogany shavings. Referring to the meeting of Rooks for judicial purposes, Mr. Strickland states that he once saw a Rook tried in this way, and ultimately killed by the rest.

SONGS OF BIRDS AND SEASONS OF THE DAY.

Although nearly half a century has elapsed since the following observations were communicated to the Royal Society by Dr. Jenner, their expressive character is as charming as ever, and their accuracy as valuable:—

"There is a beautiful propriety in the order in which Singing Birds fill up the day with their pleasing harmony. The accordance between their songs, and the aspect of nature at the successive periods of the day at which they sing, is so remarkable that one cannot but suppose it to be the result of benevolent design.

"From the *Robin* (not the *Lark*, as has been generally imagined), as soon as twilight has drawn its imperceptible line between night and day, begins his artless song. How sweetly does this harmonize with the soft dawning of the day! He goes on till the twinkling sunbeams begin to tell him that his notes no longer accord with the rising sun. Up starts the *Lark*, and with him a variety of sprightly songsters, whose lively notes are in perfect correspondence with the gaiety of the morning. The general warbling continues, with now and then an interruption by the transient croak of the *Raven*, the scream of the *Jay*, or the pert chattering of the *Daw*. The *Nightingale*, unwearied by the vocal exertions of the night, joins his inferiors in sound in the general harmony. The *Thrush* is wisely placed on the summit of some lofty tree, that its piercing notes may be softened by distance before it reaches the ear, while the mellow *Blackbird* seeks the lower branches.

"Should the sun, having been eclipsed by a cloud, shine forth with fresh effulgence, how frequently we see the *Goldfinch* perch on some blossomed bough, and hear his song poured forth in a strain peculiarly energetic; while the sun, full shining on his beautiful plumes, displays his golden wings and crimson crest to charming advantage. Indeed, a burst of sunshine in a cloudy day, or after a heavy shower, seems always to wake up a new gladness in the little musicians, and invite them to an answering burst of minstrelsy.

"As evening advances, the performers gradually retire, and the concert softly dies away. At sunset the *Robin* again sends up his twilight song, till the still more serene hour of night sends him to his bower of rest. And now, in unison with the darkened earth and sky, no sooner is the voice of the *Robin* hushed, than the *Owl* sends forth his slow and solemn tones, well adapted to the serious hour."

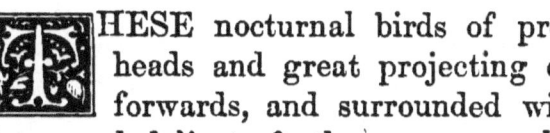HESE nocturnal birds of prey have large heads and great projecting eyes, directing forwards, and surrounded with a circle of loose and delicate feathers, more or less developed, according to the nocturnal or comparatively diurnal habits of the species. The position of the eyes, giving a particular fulness and breadth to the head, has gained for the Owl the intellectual character so universally awarded to it. The concave facial disc of feathers with which they are surrounded materially aids vision by concentrating the rays of light to an intensity better suited to the opacity of the medium in which power is required to be exercised. "They may be compared," says Mr. Yarrell, "to a person near-sighted, who sees objects with superior magnitude and brilliancy when within the prescribed limits of his natural powers of vision, from the increased angle these objects subtend." Their beaks are completely curved, or raptorial; they have the power of turning the outer toe either

backwards or forwards; they fly weakly, and near the ground; but, from their soft plumage, stealthily, stretching out their hind legs that they may balance their large and heavy heads. Their sense of hearing is very acute: they not only look, but listen for prey.

The Owl is a bird of mystery and gloom, and a special favourite with plaintive poets. We find him with Ariel:—

> "There I couch when Owls do cry."

He figures in the nursery rhyme of "Cock Robin." In reply to "Who dug his grave?"—

> "I, says the Owl, with my little shovel—
> I dug his grave."

He hoots over graves, and his dismal note adds to the terror of darkness:—

> "'Tis the middle of night by the castle clock,
> And the Owls have awakened the crowing cock;
>   Tu-whit! tu-whoo!
> And hark again the crowing cock,
>   How drowsily it crew.

> . . . .

> "When blood is nipt, and ways be foul,
> Then nightly sings the staring Owl,
>   Tu-whoo!
> Tu-whit! tu-whoo! a *merry note*,
> While greasy Joan doth keel the pot!"

Titania sings of

> "The clamorous Owl, that nightly hoots and wonders
> At our quaint spirits."

Bishop Hall has this "Occasional Meditation" upon the sight of an Owl in the twilight :—" What a strange melancholic life doth this creature lead; to hide her head all the day long in an ivy-bush, and at night, when all other birds are at rest, to fly abroad and vent her harsh notes. I know not why the ancients have *sacred* this bird to wisdom, except it be for her safe closeness and singular perspicuity; that when other domestrial and airy creatures are blind, she only hath insured light to discern the least objects for her own advantage." We may here note that Linnæus, with many other naturalists and antiquaries, have supposed the Horned Owl to have been the bird of Minerva; but Blumenbach has shown, from the ancient works of Grecian art, that it was not this, but rather some smooth-headed species, probably the *Passerina*, or Little Owl.

The divine has, in the above passage, overstated the melancholy of the Owl; as has also the poet, who sings :—

> " From yonder ivy-mantled tower
> The moping Owl does to the moon complain
> Of such as, wandering near her secret bower,
> Molest her ancient solitary reign."

Shakspeare more accurately terms her "the mousing Owl," for her nights are spent in barns, or in hunting and devouring sparrows in the church-yard elms. "Moping, indeed!" says a pleasing observer. "So far from this, she is a sprightly, active ranger of the night, who had as lief sit on a grave as a rose-bush; who is as valiant a hunter as Nimroud, chasing all sorts of game, from the dormouse to the hare and the young lamb, and devouring them, while her mate hoots to her from some picturesque ruin, and invites her, when supper is over, to return to him and her babes."

But the tricks of the Owl by night render her the terror of all other birds, great and small. In Northern Italy, persons in rustic districts which are well wooded, catch and tame an Owl, put a light chain upon her legs, and then place her on a small cross-bar on the top of a high pole, which is fixed in the earth. Half-blinded by the light, the defenceless captive has to endure patiently the jeers and insults of the dastardly tribes from the surrounding groves and thickets, who issue in clouds to scream, chirp, and flit about their enemy. Some, trusting to the swiftness of their wings, sweep close by, and peck at her feathers as they pass, and are sometimes punished by the Owl with her formidable beak for their audacity. Meanwhile, from darkened windows, sportsmen, with fowling-pieces well charged with shot, fire at the hosts of birds, wheeling, shrieking, screaming, and thickening around the Owl. All the guns are fired at once, and the grass is strewn for many yards round with the slain; while the Owl, whom they have been careful not to hit, utters a joyous whoo! whoo! at the fate of her persecutors.

Major Head thus describes the *Biscacho*, or Coquimbo, a curious species of Owl, found all over the pampas of South America:—" Like rabbits, they live in holes, which are in groups in every direction. These animals are never seen in the day, but as soon as the lower limb of the sun reaches the horizon, they are seen issuing from the holes. The Biscachos, when full-grown, are nearly as big as badgers, but their head resembles a rabbit's, except that they have large bushy whiskers. In the evening they sit outside the holes, and they all appear to be moralizing. They are the most serious-looking animals I ever saw; and even the young ones are grey-headed, wear moustachios, and look thoughtful and grave. In the daytime their holes are guarded by two little owls, which are never an instant away from their posts. As one gallops by these owls, they always stand looking at the stranger, and then at each other, moving their old-fashioned heads in a manner which is quite ridiculous, until one rushes by them, when they get the better of their dignified looks, and they both run into the Biscacho's hole."

Of all birds of prey, Owls are the most useful to man, by protecting his corn-fields, or granaried provision, from mice and numberless vermin. Yet, prejudice has perverted these birds into objects of superstition and consequent hate. The kind-hearted Mr. Waterton says:—" I wish that any little thing I could write or say might cause this bird to stand better with the world at large than it has hitherto done; but I have slender hope on this score, because old and deep-rooted prejudices are seldom overcome; and when I look back into annals of remote anti-

P

quity, I see too clearly that defamation has done its worst to ruin the whole family, in all its branches, of this poor, harmless, useful friend of mine."

The Barn Owl is common throughout Europe, known in Tartary, and rare in the United States of America. In England it is called the Barn Owl, the Church Owl, Gillihowlet, and Screech Owl; the last name is improperly applied, as it is believed not to hoot, though Sir William Jardine asserts that he has shot it in the act of hooting. To the screech superstition has annexed ideas of fatal portent; "but," says Charlotte Smith, "it has, of course, no more foreknowledge of approaching evil to man than the Lark: its cry is a signal to its absent mate."

"If," says Mr. Waterton, "this useful bird caught its food by day instead of hunting for it by night, mankind would have ocular demonstration of its utility in thinning the country of mice; and it would be protected and encouraged everywhere. It would be with us what the Ibis was with the Egyptians. When it has young, it will bring a mouse to the nest every twelve or fifteen minutes." Mr. Waterton saw his Barn Owl fly away with a rat which he had just shot; he also saw her drop perpendicularly into the water, and presently rise out of it with a fish in her claws, which she took to her nest.

Birds and quadrupeds, and even fish, are the food of Owls, according to the size of the species. Hares, partridges, grouse, and even the turkey, are attacked by the larger Horned Owls of Europe and America; while mice, shrews, small birds, and crabs suffice for the inferior strength of the smaller Owls. Mr. Yarrell states that the Short-eared Owl is the only bird of prey in which he ever found the remains of a bat.

William Bullock reports that a large Snowy Owl, wounded on the Isle of Baltoe, disgorged a young rabbit; and that one in his possession had in its stomach a sandpiper with its feathers entire. It preys on lemmings, hares, and birds, particularly the willow-grouse and ptarmigan. It is a dexterous fisher, grasping the fish with an instantaneous stroke of the foot as it sails along near the surface of the water, or sits on a stone in a shallow stream. It has been seen on the wing pursuing an American hare, making repeated strokes at the animal with its foot. In winter, when this Owl is fat, the Indians and white residents in the Fur Countries esteem it to be good eating; its flesh is delicately white. Small snakes are the common prey of this Owl during the daytime. And to show on what various kinds of food Owls subsist, Mr. Darwin states that a species that was killed among the islets of the Chonos Archipelago had its stomach full of good-sized crabs. Such are a few of the facts which attest the almost omnivorous appetite of the Owl.

The flight of the Snowy Owl is stronger and swifter than any other bird of the family; its ears are very large; its voice (says Pennant) adds horror even to the regions of Greenland by its hideous cries, resembling those of a man in deep distress. The eye is very curious, being immovably fixed in its socket, so that the bird, to view different objects, must always turn its head; and so excellently is the neck adapted to this purpose, that it can with ease turn the head round in almost a complete circle, without moving the body. The Virginian Eagle-Owl, amidst the forests of Indiana, utters a loud and sudden *Wough O! wough O!* sufficient to alarm a whole garrison; another of its nocturnal cries resembles the half-suppressed screams of a person being suffocated or throttled.

The Javanese Owl is found in the closest forests, and occasionally near villages and dwellings. Dr. Horsfield says:—"It is not, however, a favourite with the natives; various superstitious notions are also in Java associated with its visits; and it is considered in many parts of the island as portending evil." One of this species never visits the villages, but resides in the dense forests, which are the usual resort of the tiger. The natives even assert that the *Wowo-wiwi* approaches the animal with the same familiarity with which the jallack approaches the buffalo, and that it has no dread to alight on the tiger's back. Dr. Horsfield adds, that it has never been seen in confinement.

The Boobook Owl has the native name of Buckbuck, and it may be heard in Australia every night during winter, uttering a cry corresponding with that word. The note is somewhat similar to that of the European *Cuckoo,* and the colonists have given it that name. The lower order of settlers in New South Wales are led away by the idea that everything is the reverse in that country to what it is in

FRASER'S EAGLE-OWL, FROM FERNANDO PO.

England; and the *Cuckoo,* as they call this bird, singing by night, is one of the instances which they point out.

Tame Owls are described as nearly as playful, and quite as affectionate, as kittens; they will perch upon your wrist, touch your lips with their beak, and hoot to order; and they are less inclined to leave their friends than other tame birds. A writer in " Chambers's Journal " relates, that a friend lost his favourite Owl, which flew away, and was absent many days. In time, however, he came back, and resumed his habits and duties, which, for a while, went on uninterruptedly. At length, one severe autumn, he disappeared; weeks, months passed, and he returned not. One snowy night, however, as his master sat by the blazing fire, some heavy thing came bump against the shutters. "Whoo, whoo, whoo." The window was opened, and in flew the Owl, shaking the thick snow from his wings, and settling lovingly on his master's wrist, the bird's eyes dilating with delight.

The Owls at Arundel Castle have a sort of historic interest; they are kept within the circuit of the keep-tower, the most ancient and picturesque portion of the castle. Among the Australian Owls here we read of one larger than a turkey, measuring four feet across the wings when expanded. The Owl named "Lord Thurlow," from his resemblance to that Judge, is a striking specimen.

The accompanying illustration shows a fine specimen of Fraser's Eagle-Owl, brought from Fernando

Po. It is the size of an ordinary fowl; colour, very dark reddish-brown mottling; back and wings passing through all shades of the same colour into nearly white on the under parts, where the feathers are barred; bill, pale greenish; eyes, nearly black.

Among the Owls but recently described is the Masked Owl of New Holland, named from the markings of the disk of the face, somewhat grotesque; the colours are brown variegated with white. A fine specimen of the Abyssinian Owl is possessed by Mr. R. Good, of Yeovil: the bird, although quite young, is of immense size.

Lastly, the Owl is thought to be of the same sympathy or kindred likings as the Cat: a young Owl will feed well, and thrive upon fish. Cats, too, it is well known, like fish. Both the Cat and the Owl, too, feed upon mice. The sight of Owls, also, similar to that of Cats, appears to serve them best in the dark.

WEATHER-WISE ANIMALS.

WHATEVER may be the worth of weather prognostications, it is from the animal kingdom that we obtain the majority. How these creatures become so acutely sensible of the approach of particular kinds of weather is not at present well understood. That in many cases the appearance of the heavens is not the source from which their information is derived is proved by the signs of uneasiness frequently expressed by them when, as yet, the most attentive observer can detect no signs of change, and even when they are placed in such circumstances as preclude the possibility of any instruction from this quarter. For instance, Dogs, closely confined in a room, often become very drowsy and stupid before rain; and a leech, confined in a glass of water, has been found, by its rapid motions or its quiescence, to indicate the approach of wet or the return of fair weather. Probably the altered condition of the atmosphere with regard to its electricity, which generally accompanies change of weather, may so affect their constitution as to excite in them pleasurable or uneasy sensations; though man is far from insensible to atmospheric changes, as the feelings of utter listlessness which many persons experience before rain, and the aggravated severity of toothache, headache, and rheumatism abundantly testify. The Cat licking itself is a special influence of the above electric influence, which denotes the approach of rain.

Birds, as "denizens of the air," are the surest indicators of weather changes. Thus, when swallows fly high, fine weather is to be expected or continued; but when they fly low, or close to the ground, rain is almost surely approaching; for swallows follow the flies and gnats, which delight in warm strata of air. Now, as warm air is lighter, and usually moister than cold air, when the warm strata of air are high there is less chance of moisture being thrown down from them by their mixture with cold air; but when the warm and moist air is close to the surface, it is almost certain that, as the cold air flows down into it, a deposition of water will take place.

When Seagulls assemble on the land, very stormy and rainy weather is approaching. The cause of this migration to the land is the security of these birds finding food; and they may be observed at this time feeding greedily on the earth-worms and larvæ driven out of the ground by severe floods; whilst the fish on which they prey in fine weather in the sea, leave the surface, and go deeper in storms. The search after food is the principal cause why animals change their places. The different tribes of the wading birds always migrate when rain is about to take place.

There is a bird which takes its name from its apparent agency in tempests. Such is the Stormy Petrel, which name Hawkesworth, in his "Voyages," mentions the sailors give to the bird, but explains no further. Navigators meet with the Little Petrel, or Storm Finch, in every part of the ocean, diving, running on foot, or skimming over the highest waves. It seems to foresee the coming storm long ere the seamen can discover any signs of its approach. The Petrels make this known by congregating together under the wake of the vessel, as if to shelter themselves, and they thus warn the mariner of the coming danger. At night they set up a piercing cry. This usefulness of the bird to the sailor is the obvious cause of the latter having such an objection to their being killed.

Mr. Knapp, the naturalist, thus pictures gulls, describing the Petrel's action:—" They seem to repose in a common breeze, but upon the approach or during the continuation of a gale, they surround the ship, and catch up the small animals which the agitated ocean brings near the surface, or any food that may be dropped from the vessel. Whisking like an arrow through the deep valleys of the abyss, and darting away over the foaming crest of some mountain-wave, they attend the labouring barque in all her perilous course. When the storm subsides they retire to rest, and are seen no more."

Our sailors have, from very early times, called these birds "Mother Carey's Chickens," originally bestowed on them, Mr. Yarrell tells us, by Captain Cartaret's sailors, probably from some celebrated ideal hag of the above name. Mr. Yarrell adds:—
"As these birds are supposed to be seen only before stormy weather, they are not welcome visitors," a view at variance with that already suggested.

The Editor of "Notes and Queries" considers the Petrels to have been called *chickens* from their diminutive size. The largest sort, "the Giant Petrel," is "Mother Carey's *Goose;*" its length is forty inches, and it expands seven feet. The common kind are about the size of a swallow, and weigh something over an ounce; length, six inches; expansion, thirteen inches; these are Mother Carey's *chickens* (*Latham*). It should be borne in mind that our language does not restrict the term chickens to young birds of the gallinaceous class.

The Missel-bird is another bird of this kind: in Hampshire and Sussex it is called the *Storm Cock*, because it sings early in the spring, in blowing, showery weather.

Petrels, by the way, are used by the inhabitants of the Faroe Islands as lamps: they pass a wick through their bodies which, when lighted, burns a long time from the quantity of fat they contain.

The Fulmar Petrel, in Boothia, follows the whale-ships, availing itself of the labours of the fishermen by feeding on the carcases of the whales when stripped of their blubber. In return the bird is exceedingly useful to the whalers by guiding them to the places where whales are most numerous, and crowding to the spots where they first appear on the surface of the water.

Wild Geese and Ducks are unquestionably weather-wise, for their early arrival from the north in the winter portends that a severe season is approaching; because their early appearance is most likely caused by severe frost having already set in at their usual summer residence. The Rev. F. O. Morris, the well-known writer on natural history, records from Nunburnholme, Yorkshire, December 5, 1864 :—" This season, for the first time I have lived here, I have missed seeing the flocks of Wild Geese which in the autumnal months have heretofore wended their way overhead, year after year, as regularly as the dusk of the evening came on. Almost to the minute, and almost in the same exact course, they have flown over aloft from the feeding-places on the Wolds to their resting-places for the night; some, perhaps, to extensive commons, while others have turned off to the mud-banks of the Humber, whence they have returned with equal regularity in the morning.

" But this year I have seen not only not a single flock, but not even a single bird. One evening one of my daughters did indeed see a small flock of six, but even that small number only once. Whether it portends a very hard winter, or what the cause of it may be, I am utterly at a loss to know or even to guess. I quite miss this year the well-known cackle of the old gander as he has led the van of the flock that has followed him; now in a wide, now in a narrow, now in a short, now in a long wedge, over head, diverging just from the father of the family, or separating from time to time further back in the line.

"I may add, as a possible prognostication of future weather, that fieldfares have, I think, been unusually numerous this year, as last year they were the contrary. I have also remarked that swallows took their departure this year more than ordinarily in a body, very few stragglers being subsequently seen."

It will be sufficient to state that the mean temperature of January and February was below that of the same month in the preceding year, and that of March had not been so low for twenty years.

The opinion that sea-birds come to land in order to avoid an approaching storm is stated to be erroneous; and the cause assigned is, that as the fish upon which the birds prey go deep into the water during storms, the birds come to land merely on account of the greater certainty of finding food there than out at sea.

We add a few notes on Bird naturalists. The Redbreast has been called *the Naturalist's Barometer*. When on a summer evening, though it be unsettled and rainy, he sings cheerfully and sweetly on a lofty twig or housetop, it is an unerring promise of succeeding fine days. Sometimes, though the atmosphere be dry and warm, he may be seen melancholy chirping and brooding in a bush or low in a hedge; this promises the reverse. In the luxuriant forests of Brazil the Toucan may be heard rattling with his large hollow beak, as he sits on the outermost branches, calling in plaintive notes for rain.

When Mr. Loudon was at Schwetzingen, Rhenish Bavaria, in 1829, he witnessed in the post-house there for the first time what he afterwards frequently saw—an amusing application of zoological knowledge for the purpose of prognosticating the weather. Two tree-frogs were kept in a crystal jar about eighteen inches high and six inches in diameter, with a depth of three or four inches of water at the bottom, and a small ladder reaching to the top of the jar. On the approach of dry weather the frogs mounted the ladder, but when moisture was expected they descended into the water. These animals are of a bright green, and in their wild state climb the trees in search of insects, and make a peculiar singing noise before rain. In the jar they got no other food than now and then a fly; one of which, Mr. Loudon was assured, would serve a frog for a week, though it would eat from six to twelve flies in a day if it could get them. In catching the flies put alive into the jar the frogs displayed great adroitness.

Snails are extraordinary indicators of changes in the weather. Several years ago, Mr. Thomas, of Cincinnati, known as an accredited observer of natural phenomena, published some interesting accounts of Weather-wise Snails. They do not drink (he observes), but imbibe moisture in their bodies during rain, and exude it at regular periods afterwards. Then a certain snail first exudes the pure liquid; when this is exhausted, a light red succeeds, then a deep red, next yellow, and lastly a dark brown. The snail is very careful not to exude more of its moisture than is necessary. It is never seen abroad *except before rain*, when we find it ascending the bark of trees and getting on the leaves. The tree-snail is also seen ascending the stems of plants *two days before rain:* if it be a long and hard rain they get on the sheltered side of the leaf, but if a short rain the outside of the leaf. Another snail has the same habits, but differs only in colour : before rain it is yellow, and after it blue. Others show signs of rain, not only by means of exuding fluids, but by means of pores and protuberances ; and the bodies of some snails have large tubercles rising from them *before rain*. These tubercles commence showing themselves ten days previous to the fall of rain they indicate; at the end of each of these tubercles is a pore ; and at the time of the fall of rain these tubercles, with their pores opened, are stretched to their utmost to receive the water. In another kind of snail, a few days before rain appears a large and deep indentation, beginning at the head between the horns, and ending with the jointure at the shells. Other snails, a few days before the rain, crawl to the most exposed hill-side, where, if they arrive before the rain descends, they seek some crevice in the rocks, and then close the aperture of the shell with glutinous substance; this, when the rain approaches, they dissolve, and are then seen crawling about.

Our Cincinnati observer mentions three kinds of snails which move along at the rate of a mile in forty-four hours; they inhabit the most dense forests, and it is regarded as a sure indication of rain to observe them moving towards an exposed situation. Others indicate the weather not only by exuding fluids, but by the colour of the animal. After rain the snail has a very dark appearance, but it grows of a bright colour as the water is expended; whilst just before rain it is of yellowish white colour, also just before rain streaks appear from the point of the head to the jointure of the shell. These snails move at the rate of a mile in fourteen days and sixteen hours. If they are observed ascending a cliff it is a sure indication of rain: they live in the cavities of the sides of cliffs. There is also a snail which is brown, tinged with blue on the edges before rain, but black after rain: a few days before appears an indentation, which grows deeper as the rain approaches.

The leaves of trees are even good barometers: most of them for a short, light rain, will turn up so as to receive their fill of water; but for a long rain they are doubled, so as to conduct the water away. The Frog and Toad are sure indicators of rain ; for, as they do not drink water but absorb it into their bodies, they are sure to be found out at the time they expect rain. The Locust and Grasshopper are also good indicators of a storm; a few hours before rain they are to be found under the leaves of trees and in the hollow trunks.

The Mole has long been recorded as a prognosticator of change of weather, before which it becomes very active. The temperature or dryness of the air governs its motions as to the depth at which it lives or works. This is partly from its inability to bear cold or thirst, but chiefly from its being necessitated to follow its natural food, the earthworm, which always descends as the cold or drought increases. In frosty weather both worms and moles are deeper in the ground than at other times; and both seem to be sensible of an approaching change to warmer weather before there are any perceptible signs of it in the atmosphere. When it is observed, therefore, that Moles are casting hills through openings in the frozen turf or through a thin covering of snow, a change to open weather may be shortly expected. The cause of this appears to be—the natural heat of the earth being for a time pent in by the frozen surface accumulates below it; first incites to action the animals, thaws the frozen surface, and at length escapes into the air, which is warm, and softens; and if not counterbalanced by a greater degree of cold in the atmosphere brings about a change, such as from frosty to mild weather. The Mole is most active and casts up most earth immediately before rain, and in the winter before a thaw, because at those times the worms and insects begin to be in motion, and approach the surface.

Forster, the indefatigable meteorologist, has assembled some curious observations on certain animals, who, by some peculiar sensibility to electrical or other atmospheric influence, often indicate changes of the weather by their peculiar motions and habits. Thus :—

Ants.—An universal bustle and activity observed in ant-hills may be generally regarded as a sign of rain : the Ants frequently appear all in motion together, and carry their eggs about from place to place. This is remarked by Virgil, Pliny, and others.

Asses.—When donkeys bray more than ordinarily, especially should they shake their ears, as if uneasy, it is said to predict rain, and particularly showers. Forster noticed that in showery weather a donkey brayed before every shower, and generally some minutes before the rain fell, as if some electrical influence, produced by the concentrating power of the approaching rain-cloud, caused a tickling in the wind-pipe of the animal just before the shower came on. Whatever this electric state of the air preceding a shower may be, it seems to be the same that causes in other animals some peculiar sensations, which makes the peacock squall, the pintado call "come back," &c. An expressive adage says :—

"When that the ass begins to bray,
Be sure we shall have rain that day."

Haymakers may derive useful admonitions from the braying of the ass : thus the proverb :—

> "Be sure to cock your hay and corn
> When the old donkey blows his horn."

Bats flitting about late in the evening in spring and autumn foretel a fine day on the morrow; as do Dorbeetles and some other insects. On the contrary, when Bats return soon to their hiding-places, and send forth loud cries, bad weather may be expected.

Beetles flying about late in the evening often foretel a fine day on the morrow.

Butterflies, when they appear early, are sometimes forerunners of fine weather. Moths and Sphinxes also foretel fine weather when they are common in the evening.

Cats, when they "wash their faces," or when they seem sleepy and dull, foretel rain.

Chickens, when they pick up small stones and pebbles, and are more noisy than usual, afford a sign of rain; as do fowls rubbing in the dust, and clapping their wings; but this applies to several kinds of fowls, as well as to the gallinaceous kinds. Cocks, when they crow at unwonted hours, often foretel rain; when they crow all day, in summer particularly, a change to rain frequently follows.

Cranes were said of old to foretel rain when they retreated to the valleys, and returned from their aërial flight. The high flight of cranes in silence indicates fine weather.

Dolphins as well as *Porpoises*, when they come about a ship, and sport and gambol on the surface of the water, betoken a storm.

Dogs, before rain, grow sleepy and dull, lie drowsily before the fire, and are not easily aroused. They also often eat grass, which indicates that their stomachs, like ours, are apt to be disturbed before change of weather. It is also said to be a sign of change of weather when Dogs howl and bark much in the night. Dogs also dig in the earth with their feet before rain, and often make deep holes in the ground.

Ducks.—The loud and clamorous quacking of Ducks, Geese, and other water-fowl, is a sign of rain; as also when they wash themselves, and flutter about in the water more than usual. Virgil has well described all these habits of aquatic birds.

Fieldfares, when they arrive early, and in great numbers, in autumn, foreshow a hard winter, which has probably set in in the regions from which they have come.

Fishes, when they bite more readily, and gambol near the surface of streams or pools, foreshow rain.

Flies, and various sorts of insects, become more troublesome, and sting and bite more than usual, before, as well as in the intervals of rainy weather, particularly in autumn.

Frogs, by their clamorous croaking, indicate rainy weather, as does likewise their coming about in great numbers in the evening; but this last sign applies more obviously to toads.

Geese washing, or taking wing with a clamorous noise, and flying to the water, portend rain.

Gnats afford several indications. When they fly in a vortex in the beams of the setting sun they forebode fair weather; when they frisk about more widely in the open air at eventide they foreshow heat; and when they assemble under trees, and bite more than usual, they indicate rain.

Hogs, when they shake the stalks of corn, and spoil them, often indicate rain. When they run squeaking about, and jerk up their heads, windy weather is about to commence; hence the Wiltshire proverb, that "Pigs can see the wind."

Horses foretel the coming of rain by starting more than ordinarily, and by restlessness on the road.

Jackdaws are unusually clamorous before rain, as are also *Starlings*. Sometimes before change of weather the daws make a great noise in the chamber wherein they build.

Kine (cattle) are said to foreshow rain when they lick their fore-feet, or lie on their right side. Some say oxen licking themselves against the hair is a sign of wet.

Kites, when they soar very high in the air, denote fair weather, as do also *Larks*.

Magpies, in windy weather, often fly in small flocks of three or four together, uttering a strong harsh cry.

Mice when they squeak much, and gambol in the house, foretel a change of weather, and often rain.

Owls.—When an owl hoots or screeches, sitting on the top of a house, or by the side of a window, it is said to foretel death. "The fact," says Forster, "seems to be this: the Owl, as Virgil justly observes, is more noisy at the change of weather, and as it often happens that patients with lingering diseases die at the change of weather, so the Owl seems, by a mistaken association of ideas, to forebode the calamity.

Peacocks squalling by night often foretel a rainy day. Forster adds, "This prognostic does not often fail; and the indication is made more certain by the crowing of Cocks all day, the braying of the Donkey, the low flight of Swallows, the aching of rheumatic persons, and by the frequent appearance of spiders on the walls of the house."

Pigeons.—It is a sign of rain when Pigeons return slowly to the dove-houses before the usual time of day.

Ravens, when observed early in the morning, at a great height in the air, soaring round and round, and uttering a hoarse, croaking sound, indicate that the day will be fine. On the contrary, this bird affords us a sign of coming rain by another sort of cry; the difference between these two voices being more easily learned from nature than described. The Raven frequenting the shore and dipping himself in the water is also a sign of rain.

Redbreasts, when they, with more than usual familiarity, lodge on our window-frames, and peck against the glass with their bills, indicate severe weather, of which they have a presentiment, which brings them nearer to the habitations of man.

Rooks gathering together, and returning home from their pastures early, and at unwonted hours, forebode rain. When Rooks whirl round in the air rapidly, and come down in small flocks, making a roaring noise with their wings, rough weather invariably follows. On the contrary, when Rooks are very noisy about their trees, and fly about as if rejoicing, Virgil assures us they foresee a return of fine weather, and an end of the showers.

Spiders, when seen crawling on the walls more than usual, indicate rain. "This prognostic," says Forster, "seldom fails. I have noticed it for many years, particularly in winter, but more or less at all times of the year. In summer the quantity of webs of the garden spiders denote fair weather."

Swallows, in fine and settled weather, fly higher in the air than they do just before or during a showery or rainy time. Then, also, Swallows flying low, and skimming over the surface of a meadow where there is tolerably long grass, frequently stop, and hang about the blades, as if they were gathering insects lodged there.

Swans, when they fly against the wind, portend rain, a sign frequently fulfilled.

Toads, when they come from their holes in an unusual number in the evening, although the ground be still dry, foreshow the coming rain, which will generally fall more or less during the night.

Urchins of the Sea, a sort of fish, when they

thrust themselves into the mud, and try to cover their bodies with sand, foreshow a storm.

Vultures, when they scent carrion at a great distance, indicate that state of the atmosphere which is favourable to the perception of smells, and this often forebodes rain.

Willow Wrens are frequently seen, in mild and still rainy weather, flitting about the willows, pines, and other trees, in quest of insects.

Woodcocks appear in autumn earlier, and in greater numbers, previous to severe winters; as do Snipes and other winter birds.

Worms come forth more abundantly before rain, as do snails, slugs, and almost all limaceous animals.

Some birds build their nests weather-proof, as ascertained by careful observation of Mr. M. W. B. Thomas, of Cincinnatti, Ohio. Thus, when a pair of migratory birds have arrived in the spring, they prepare to build their nest, making a careful reconnaissance of the place, and observing the character of the season. If it be a windy one, they thatch the straw and leaves on the inside of the nest, between the twigs and the lining; if it be very windy, they get pliant twigs, and bind the nest firmly to the limb of the tree, securing all the small twigs with their saliva. If they fear the approach of a rainy season, they build their nests so as to be sheltered from the weather; but if a pleasant one, they build in a fair open place, without taking any of these extra precautions.

Of all writers, Dr. Darwin has given us the most

correct account of the "Signs of Rain," in a poetical description of the approach of foul weather, as follows. This passage has been often quoted, but, perhaps, never exceeded in the accuracy of its phenomenal observation :—

> "The hollow winds begin to blow;
> The clouds look black, the glass is low;
> The soot falls down, the spaniels sleep;
> And spiders from their cobwebs peep.
> Last night the sun went pale to bed;
> The moon in haloes hid her head;
> The boding shepherd heaves a sigh,
> For, see, a rainbow spans the sky.
> The walls are damp, the ditches smell,
> Clos'd is the light red pimpernel.
> Hark! how the chairs and tables crack,
> Old Betty's joints are on the rack;
> Her corns with shooting pains torment her,
> And to her bed untimely send her.
> Loud quack the ducks, the sea-fowls cry,
> The distant hills are looking nigh.
> How restless are the snorting swine!
> The busy flies disturb the kine.
> Low o'er the grass the swallow wings,
> The cricket, too, how sharp he sings!
> Puss on the hearth, with velvet paws,
> Sits wiping o'er her whisker'd jaws.
> The smoke from chimneys right ascends;
> Then spreading back, to earth it bends.
> The wind unsteady veers around,
> Or settling in the South is found.
> Through the clear stream the fishes rise,
> And nimbly catch th' incautious flies.
> The glowworms num'rous, clear, and bright,
> Illum'd the dewy hill last night.

> At dusk, the squalid toad was seen,
> Like quadruped, stalk o'er the green.
> The whirling wind the dust obeys,
> And in the rapid eddy plays.
> The frog has chang'd his yellow vest,
> And in a russet coat is drest.
> The sky is green, the air is still,
> The mellow blackbird's voice is shrill.
> The dog, so altered is his taste,
> Quits mutton-bones on grass to feast.
> Behold the rooks, how odd their flight,
> They imitate the gliding kite,
> And seem precipitate to fall,
> As if they felt the piercing ball.
> The tender colts on banks do lie,
> Nor heed the traveller passing by.
> In fiery red the sun doth rise,
> Then wades through clouds to mount the skies.
> ' 'Twill surely rain, we see 't with sorrow,
> No working in the fields to-morrow.' "

The Shepherd of Banbury says:—" The surest and most certain sign of rain is taken from Bees, which are more incommoded by rain than almost any other creatures; and, therefore, as soon as the air begins to grow heavy, and the vapours to condense, they will not fly from their hives, but either remain in them all day, or else fly but to a small distance." Yet Bees are not always right in their prognostics, for Réaumur witnessed a swarm which, after leaving the hive at half-past one o'clock, were overtaken by a heavy shower at three.

FISH-TALK.

"MAN favours wonders;" and this delight is almost endlessly exemplified in the stories of strange Fishes—of preternatural size and odd forms, which are to be found in their early history. In our present Talk we do not aim at reassembling these olden tales, but propose rather to glance at recent accessions to our acquaintance with the study of Fish-life, and a few modern instances of the class of wonders.

Fishes, like all other animals, have a very delicate sense of the equilibrial position of their bodies. They endeavour to counteract all change in their position by means of movements partly voluntary and partly instinctive. These latter appear in a very remarkable manner in the eye; and they are so constant and evident in fishes while alive, that their absence is sufficient to indicate the death of the animal. The equilibrium of the fish, its horizontal position, with the back upwards, depends solely on the action of the fins, and principally that of the vertical fins. The swimming-bladder may enable a fish to increase or diminish its specific gravity. By compressing the air contained in it, the fish descends in the water; it rises by releasing the muscles which produced the compression. By compressing more or less the posterior or anterior portion of the bladder, the animal, at pleasure, can make the anterior or posterior half of its body lighter; it can also assume an oblique position, which permits an ascending or descending movement in the water.

There is a small fish found in the rivers of the Burmese Empire, which, on being taken out of the water, has the power of blowing itself up to the shape of a small round ball, but its original shape is resumed as soon as it is returned to the river.

Mr. St. John, in his "Tour in Eastern Lanarkshire," gives some curious instances of fish changing colour, which takes place with surprising rapidity. Put a living black burn Trout into a white basin of water, and it becomes, within half an hour, of a light colour. Keep the fish living in a white jar for some days, and it becomes absolutely white; but put it into a dark-coloured or black vessel, and although on first being placed there the white-coloured fish shows most conspicuously on the black ground, in a quarter of an hour it becomes as dark-coloured as the bottom of the jar, and consequently difficult to be seen. No doubt this facility of adapting its colour to the bottom of the water in which it lives, is of the greatest service to the fish in protecting it from its numerous enemies. All anglers must have observed, that in every stream the Trout are very much of the same colour as the gravel or sand on which they live: whether this change of colour is a voluntary or involuntary act on the part of the fish, the scientific must determine.

Anglers of our time have proved that Tench croak like frogs; Herrings cry like mice; Gurnards grunt like hogs; and some say the Gurnard makes a noise like a cuckoo, from which he takes one of his country names. The Maigre, a large sea-fish, when swimming in shoals, utters a grunting or piercing noise, that may be heard from a depth of twenty fathoms.

M. Dufossé asserts that facts prove that nature has not refused to all fishes the power of expressing their instinctive sensations by sounds, but has not conferred on them the unity of mechanism in the formation of sonorous vibrations as in other classes of vertebrated animals. Some fishes, he says, are able to emit musical tones, engendered by a mechanism in which the muscular vibration is the principal motive power; others possess the faculty of making blowing sounds, like those of certain reptiles; and others can produce the creaking noise resembling that of many insects. These phenomena M. Dufossé has named "Fish-noise."

The River Plate swarms with fish, and is the *habitat* of one possessed of a very sonorous voice, like that found in the River Borneo—the account of which is quoted by Dr. Buist from the Journal of the Samarang; and there is similar testimony of a loud piscatory chorus being heard on board H.M.S. Eagle, anchored, in 1845-6, about three miles from Monte Video, during the night.

That fishes hear has been doubted, although John Hunter was of this opinion, and has been followed by many observers. When standing beside a person angling, how often is the request made not to make a noise, as that would *alarm* the fish. On the other hand, the Chinese drive the fish up to that part of the river where their nets are ready to capture them by loud yells and shouts, and the sound of gongs; but old Æsop writes of a fisherman who caught no fish because he alarmed them by playing on his flute while fishing. In Germany the Shad is taken by means of nets, to which bows of wood, hung with a number of little bells, are attached in such a manner as to chime in harmony when the nets are moved. The Shad, when once attracted by the sound, will not attempt to escape while the bells continue to ring. Ælian says the Shad is allured by castanets. Macdiarmid, who declares that fishes hear as well as see, relates that an old Codfish, the patriarch of the celebrated fish-pond at Logan, "answered to his name; and not only drew near, but turned up his snout most beseechingly when he heard the monosyllable 'Tom;' and that he evidently could distinguish the voice of the fisherman who superintended the pond, and fed the fish, from that of any other fisherman." In the "Kaleidoscope" mention is made of three Trout in a pond near the powder-mills at Faversham, who were so tame as to come at the call of the person accustomed to feed them. Izaak Walton tells of a Carp coming to a certain part of a pond to be fed "at the ringing of a bell, or the beating of a drum;" and Sir John Hawkins was assured by a clergyman, a friend of his, that at the Abbey of St. Bernard, near Antwerp, he saw a Carp come to the edge of the water to be fed, at the whistle of the person who fed it. The Carp at Fontainebleau, inhabiting the lake adjoining the Imperial Palace, are of great size, and manifest a curious instinct. A Correspondent of the "Athenæum" remarks:—

"Enjoying entire immunity from all angling arts and lures, the Fontainebleau Carp live a life of great enjoyment, marred only, we imagine, by their immense numbers causing the supply of food to be somewhat below their requirements. It is not, however, very easy to define what a Carp's requirements in the form of pabulum are, as he is a voracious member of the ichthyological family, eating whenever he has an opportunity until absolutely surfeited. His favourite food consists of vegetable substances masticated by means of flat striated teeth, which work with a millstone kind of motion against a singular process of the lower part of the skull covered with horny plates. When this fish obtains an abundant supply of food it grows to an enormous size. Several continental rivers and lakes are very congenial to Carp, and especially the Oder, where this fish occasionally attains the enormous weight of 60 lb. It is not probable that any Carp in the lake at Fontainebleau are so large as this; but there are certainly many weighing 50 lb., patriarchs of their kind, which, though olive-hued in their tender years, are now white with age. That the great size of these fish is due to ample feeding is, we think, evident, and, as we shall see presently, it is the large fish that are the best fed. During many years the feeding of the Carp at Fontainebleau has been a favourite Court pastime. But it is from the visitors who frequent Fontainebleau during a great part of the year that the Carp receive their most bountiful rations. For big Carp have an enormous swallow, soft penny rolls being mere mouthfuls, bolted with ostrich-like celerity. So to prevent the immediate disappearance of these *bonnes bouches,* bread, in the form of larger balls than the most capacious Carp can take into his gullet, is baked until it becomes as hard as biscuit, and with these balls the Carp are regailed. Throw one into the lake, and you will quickly have an idea of the enormous Carp population it contains. For no sooner does the bread touch the water than it is surrounded by hundreds of these fish, which dart to it from all sides. And now, if you look attentively, you will witness a curious display of instinct, which might almost take a higher name. Conscious, apparently, of their inability to crush these extremely hard balls, the Carp combine with surprising unanimity to push them to that part of the lake with their noses where it is bounded by a wall, and when there they butt at them, until at last their repeated blows and the softening effect of the water causes them to yield and open. And now you will see another curious sight. While shoals of Carp have been pounding away at the bread-balls, preparing them for being swallowed, some dozen monsters hover round, indifferent, apparently, to what is passing. But not so, for no sooner is the bread ready for eating, than two or three of these giants, but more generally one—the tyrant, probably, of the lake—rush to the prize, cleaving the shoals of smaller Carp, and shouldering them to the right and left, seize the bread with open jaws, between which it quickly disappears."

Some of the finest and oldest Carp are found in the windings of the Spree, in the tavern-gardens of Charlottenburg, the great resort of strollers from Berlin. Visitors are in the habit of feeding them with bread, and collect them together by ringing a bell, at the sound of which shoals of the fish may be seen popping their noses upwards from the water.

The affection of fishes has only been properly understood of late years. It might be supposed that little natural affection existed in this cold-blooded race; and, in fact, fishes constantly devour their own eggs, and, at a later period, their own young, without compunction or discrimination. Some few species bear their eggs about with them until hatched. This was long thought to be the utmost extent of care which fishes lavished on their young; but Dr. Hancock has stepped in to rescue at least one species from this unmerited charge. "It is asserted," he says, "by naturalists, that no fishes are known to take any care of their offspring. Both species of *Hassar* mentioned below, however, make a regular nest, in which they lay their eggs in a flattened cluster, and cover them over most carefully. Their care does not end here; they remain by the side of the nest till the spawn is hatched, with as much solicitude as a hen guards her eggs, both the male and female Hassar, for they are monogamous, steadily watching the spawn and courageously attacking the assailant. Hence the negroes frequently take them by putting their hands into the water close to the nest, on agitating which the male Hassar springs furiously at them, and is thus captured. The *roundhead* forms its nest of grass, the *flathead* of leaves. Both, at certain seasons, burrow in the bank. They lay their eggs only in wet weather. I have been surprised to observe the sudden appearance of numerous nests in a morning after rain occurs, the spot being indicated by a bunch of .froth which appears on the surface of the water over the nest. Below this are the eggs, placed on a bunch of fallen leaves or grass, which they cut and collect together. By what means this is effected seems rather mysterious, as the species are destitute of cutting-teeth. It may, possibly, be by the use of their arms, which form the first ray of the pectoral fin."

There is another operation by fishes, which seems to require almost equal experience. Professor Agassiz, while collecting insects along the shores of Lake Sebago, in Maine, observed a couple of Catfish, which, at his approach, left the shore suddenly, and returned to the deeper water. Examining the place which the fishes had left, he discovered a *nest* among the water-plants, with a number of little tadpoles. In a few moments the two fishes returned, looking anxiously towards the nest, and approached within six or eight feet of where Professor Agassiz stood. They were evidently not in search of food, and he became convinced that they were seeking the protection of their young. Large stones, thrown repeatedly into the middle of the nest after the fishes had returned to it, only frightened them away for a brief period, and they returned to the spot within ten or fifteen minutes. This was repeated four or five times with the same result. This negatives the assertion made by some naturalists—that no fishes are known to take any care of their offspring.

But affection is scarcely to be looked for where the offspring is so very numerous as to put all attempts at even recognising them out of the question. How could the fondest mother love 100,000 little ones at once? Yet the number is far exceeded by some of the matrons of the deep. Petit found 300,000 eggs in a single carp; Lenwenhoeck 9,000,000 in a single cod; Mr. Harmer found in a sole 100,000; in a tench 300,000; in a mackerel 500,000; and in a flounder 1,357,000.* M. Rousseau disburthened a pike of 160,000, and a sturgeon of 1,567,000, while from this latter class has been gotten 119 pounds weight of eggs, which, at the rate of 7 to a grain, would give a total amount of 7,653,200 eggs! If all these came to maturity the world would be in a short time nothing but fish: means, however, amply sufficient to keep down this unwelcome superabundance have been provided. Fish themselves, men, birds, other marine animals,

---

* A tench was brought to Mr. Harmer so full of spawn that the skin was burst by a slight knock, and many thousands of the eggs were lost; yet even after this misfortune he found the remainder to amount to 383,252! Of other marine animals, which he includes under the general term fish, the fecundity, though sufficiently great, is by no means enormous. A lobster yielded 7,227 eggs; a prawn 3,806; and a shrimp 3,057. See Mr. Harmer's paper, "Philosophical Transactions," 1767.

to say nothing of the dispersions produced by storms and currents, the destruction consequent on their being thrown on the beach and left there to dry up, all combine to diminish this excessive supply over demand. Yet, on the other hand (so wonderfully are all the contrivances of nature so harmonized and balanced), one of these apparent modes of destruction becomes an actual means of extending the species. The eggs of the pike, barbel, and many other fish, says M. Virey, are rendered indigestible by an acid oil which they contain, and in consequence of which they are passed in the same condition as they were swallowed; the result of which is, that being taken in by ducks, grebes, or other water-fowls, they are thus transported to situations, such as inland lakes, which otherwise they could never have attained; and in this way only can we account for the fact, now well ascertained, that several lakes in the Alps, formed by the thawing of the glaciers, are now abundantly stocked with excellent fish.

Little fishes are ordinarily the food of larger marine animals; but a remarkable exception occurs in the case of the larger Medusæ, which are stated in various works to prey upon fishes for sustenance. Mr. Peach, the naturalist, has, however, by observations at Peterhead, in Aberdeenshire, thus corrected this statement. He observed several small fishes playing round the larger Medusæ in the harbour and bay. When alarmed, they would rush under the umbrella, and remain sheltered in its large folds till the danger had passed, when they would emerge, and sport and play about their sheltering friend. When beneath the umbrella they lay so close that they were frequently taken into a bucket with the Medusæ. They proved to be young whitings, varying from 1½ to 2 inches long. These little creatures, so far from becoming the prey of the Medusæ, experienced from them protection; and, morever, they preferred the *stinging* one. In no instance did Mr. Peach see a fish in the stomach of the Medusæ, but all could liberate themselves when they pleased. In one case, Mr. Peach witnessed a small whiting, in the first instance chased by a single young pollack, whose assault the little fellow easily evaded by dodging about; but the chaser being joined by others, the whiting was driven from its imperfect shelter, and after being much bitten and dashed about by its assailants, became at length completely exhausted, and lay to all appearance dead. Recovering, however, after action, it swam slowly to the Medusæ, and took refuge as before; but its movements being soon observed, it was again attacked, after a very brief respite, driven into open water, and speedily despatched.

Fishes appear to execute annually two great migrations. By one of these shiftings they forsake the deep water for a time, and approach the shallow shores, and by the other they return to their more concealed haunts. These movements are connected with the purposes of spawning, the fry requiring to come into life, and to spend a certain portion of their youth in situations different from those which are suited to the period of maturity. It is in obedience to these arrangements that the Cod and Haddock, the Mackerel, and others, annually leave the deeper and less accessible parts of the ocean, the region of the zoophytic tribes, and deposit their spawn within that zone of marine vegetation which fringes our coasts, extending from near the high-water mark of neap-tides to a short distance beyond the low-water mark of spring-tides. Amidst the shelter in this region afforded by the groves of arborescent fuci, the young fish were wont in comfort to spend their infancy, but since these plants have been so frequently cut down to procure materials for the manufacture of kelp, and the requisite protection withdrawn, the fisheries have greatly suffered. Many species of fish, as the Salmon, Smelt, and others, in forsaking the deep water, and approaching a suitable spawning station, leave the sea altogether for a time, ascend the rivers and their tributary streams, and, having deposited their eggs, return again to their usual haunts. Even a certain species of fish, inhabiting lakes, as the Roach, betake themselves to the tributary streams, as the most suitable places for spawning.

The Goramy, of India, are stated by General Hardwicke to watch most actively the margins of the spot which they select and prepare for depositing their spawn, driving away with violence every other fish which approaches their cover. The General adds that from the time he first noticed this circumstance about one month had elapsed, when one day he saw numerous minute fishes close to the margin of the grass, on the outer side of which the parent fishes continued to pass to and fro.

There is a species of Grampus from two to three tons weight, and about sixteen feet in length, that amuses itself with jumping, or rather springing its ponderous body entirely out of the water, in a vertical position, and falling upon its back. This effort of so large a fish is almost incredible, and informs us how surprisingly great the power of muscle must be in this class of animal. A Correspondent writes to the "United Service Journal":—"I have seen them spring out of the water within ten yards of the ship's side, generally in the evening, after having swam all the former part of the day in the ship's wake, or on either quarter. When several of these fish take it into their heads to 'dance a hornpipe,' as the sailors term their gambols, at the distance of half a mile, they, especially at or just after sundown, may easily be mistaken for the sharp points of rocks sticking up out of the water, and the splashing and foam they make and produce have the appearance of the action of waves upon rocks. An officer of the navy informed me that, after sunset, when near the equator, he was not a little alarmed and surprised at the cry of 'rocks on the starboard bow!' Looking forward, he indistinctly saw objects which he and all on board took to be pinnacles of several rocks of a black and white colour. In a short time, however, he discovered this formidable danger to be nothing more than a company of dancing Grampuses with white bellies. As one disappeared, another rose; so that there were at least five or six constantly above the surface."

Captain Owen relates that "the Bonita has the power of throwing itself out of the water to an almost incredible distance when in pursuit of its prey, the Flying Fish; and, the day previous to our arrival at Mozambique, one of these fish rose close under our bow, and passed under the vessel's side, and struck with such force against the poop, that, had any one received the blow, it must have been fatal. Stunned by the violence of the contact, it fell motionless at the helmsman's feet; but, soon recovering, its struggles were so furious that it became necessary to inflict several blows with an axe before it could be approached with safety. The greatest elevation it attained above the surface of water was eighteen feet, and the length of the leap, had no opposition occurred, would have exceeded 180."

Of winged or Flying Fish we find this extravagant account in a philosophical romance, entitled, "Telliamed," by M. Maillet, an ingenious Frenchman, of the days of Louis XV.:—

He believed, like Lamarck, that the whole family of birds had existed one time as fishes, which, on being thrown ashore by the waves, had got feathers by accident; and that men themselves are but the descendants of a tribe of sea-monsters, who, tiring of their proper element, crawled upon the beach one sunny morning, and, taking a fancy to the land, forgot to return. The account is as amusing as a fairy tale. "Winged or Flying Fish," says Maillet, "stimulated by the desire of prey, or the fear of death, or pushed near the shore by the billows, have fallen among the reeds or herbage, whence it was not possible for them to resume their flight to the sea, by means of which they had contracted their first facility of flying. Then their fins, being no longer bathed in the seawater, were split and became warped by their dryness. While they found among the reeds and herbage among which they fell many aliments to support them, the vessels of their fins being separated, were lengthened, or clothed with beards, or, to speak more justly, the membranes which before kept them adherent to each other were metamorphosed. The beard formed of these warped membranes was lengthened. The skin of these animals was insensibly covered with a down of the same colour with the skin, and this down gradually increased. The little wings they had under their belly, and which, like their wings, helped them to walk into the sea, became feet, and helped them to walk on the land. There were also other small changes in their figure. The beak and neck of some were lengthened, and of others shortened. The conformity, however, of the first figure subsists in the whole, and it will be always easy to know it. Examine all the species of fowl, even those of the Indies, those which are tufted or not, those whose feathers are reversed—such as we see at Damietta, that is to say, whose plumage runs from the tail to the head—and you will see fine species of fish quite similar, scaly or without scales. All species of Parrots, whose plumages are different, the rarest and most singular marked birds, are, conformable to fact, painted, like them, black, brown, grey, yellow, green, red, violet colour, and those of gold and azure; and all this precisely in the same parts, where the plumages of these birds are diversified in so curious a manner."

The Jaculator Fish, of Java, has been called "a sporting fish," from the precision with which it takes aim at its prey. In 1828 Mr. Mitchell saw several of these fishes in the possession of a Javanese chief; and here is the account of the curious manner in which these Jaculators were employed. They were placed in a small circular pond, from the centre of which projected a pole upwards of two feet in height. At the top of the pole were inserted small pieces of wood, sharp-pointed, and on each of these were placed insects of the beetle tribe. When the slaves had placed the beetles, the fish came out of their holes, and swam round the pond. One of them came to the surface of the water, rested there, and after steadily fixing its eyes for some time on an insect, it discharged from its mouth a small quantity of watery fluid, with such force, and precision of aim, as to strike it off the twig into the water, and in an instant swallowed it. After this, another fish came, and performed a similar feat, and was followed by the others, until they had secured all the insects. If a fish failed in bringing down its prey at the first shot, it swam round the pond till it came opposite the same object, and fired again. In one instance, a fish returned three times to the attack before it secured its prey; but in general the fish seemed very expert gunners, bringing down the beetle at the first shot. The fish, in a state of nature, frequents the shores and sides of the rivers in search of food. When it spies a fly sitting on the plants that grow on shallow water, it swims on to the distance of five or six feet from them, and then, with surprising dexterity, it ejects out of its tubular mouth a single drop of water, which rarely fails to strike the fly into the sea, where it soon becomes its prey.

Curious fish, in great numbers, may be seen in the Harbour of Port Royal, Jamaica, on the surface of the water, and are ranked among the peculiarities of the place. They are the Guardo, or Guard-Fish; the Jack (Sword-Fish); and the Ballahou. The Jack is the largest, and appears to be always at war with the two others; it is armed with formidable teeth; it basks on the surface of the water during the heat of the day, in a sort of indolent, unguarded state; but this is assumed, the better to ensnare the other fish, and to catch the floating bodies that may happen to pass near it; for the moment anything is thrown into the sea from the ship, the Jack darts with the rapidity of lightning upon it, and seizing it as quickly, retreats. This Warrior-fish possesses a foresight or instinctive quality which we see sometimes exemplified in different animals, almost amounting to second reason, such as the sagacity it displays in avoiding the hook when baited; although extremely voracious, it seems aware of the lure held out for its destruction, and avoids it with as much cunning as the generality of fishes show eagerness to devour it. The situation it takes, immediately in the wake of the ship at anchor, is another instance of its sagacity; as whatever is thrown overboard passes astern, where the fish is ever on the alert for the articles thrown over. No other fish of equal size dare approach. The Jack is, however, sometimes enticed with the bait; but he is more frequently struck with a barbed lance, or entrapped in a net. The Guardo has similar habits with the Jack, but is generally beaten by him; yet the former tyrannizes with unrelenting rigour over the weaker associate, the Ballahou.

The tiger of the ocean, the Shark, is often cruising about Port Royal, but rarely injures human life. At Kingston, however, such distressing events often occur. There was a pet Shark known as "Old Tom of Port Royal;" it was fed whenever it approached any of the ships, but was at last killed by the father of a child which it had devoured. Whilst it remained here, no other of the Shark tribe dare venture on his domain; he reigned lord paramount in his watery empire, and never committed any depredation but that for which he suffered.

Attending the Shark is seen the beautiful little Pilot Fish, who, first approaching the bait, returns as if to give notice, when, immediately after, the Shark approaches to seize it. It is a curious circumstance, that this elegant little fish is seen in attendance only upon the Shark. After the Shark is hooked, the Pilot Fish still swims about, and for some time after he has been hauled on deck; it then swims very near the surface of the water. When the Shark has been hooked, and afterwards escapes, he generally returns, and renews the attack with increased ferocity, irritated often by the wound he has received.

Sharks appear to have become of late years much more numerous in Faroe, as they have also in other parts of the North Seas, especially on the coast of Norway.

The reader may, probably, have found on the sea-shore certain cases, which are fancifully called sea-purses, mermaids' purses, &c. Now, some Sharks bring forth their young alive, whilst others are enclosed in oblong semi-transparent, horny cases, at each extremity of which are two long tendrils. These cases are the above *purses*, which the parent Shark deposits near the shore in the winter months. The twisting tendrils hang to sea-weed, or other fixed bodies, to prevent the cases being washed away into deep water. Two fissures, one at each end, allow the admission of sea-water; and here the young Shark remains until it has acquired the power of taking food by the mouth, when it leaves what resembles its cradle. The young fish ultimately escapes by an opening at the end, near which the head is situated.

California has yielded an extraordinary novelty in fish history. In 1854 Mr. Jackson, while fishing in San Salita Bay, caught with a hook and line a fish of the perch family *containing living young*. These were supposed to be the prey which the fish had swallowed, but on opening the belly was found next to the back of the fish, and slightly attached to it, a long very light violet bag, so clear and transparent that there could already be distinguished through it the shape, colour, and formation of a multitude of small fish (all facsimiles of each other), with which the bag was filled. They were in all respects like the mother, and like each other; and there cannot remain a single doubt that these young were the offspring of the fish from whose body they were taken; and that this species of fish gives birth to her young alive and perfectly formed, and adapted to seek its own livelihood in the water. Professor Agassiz has confirmed the truth of this extraordinary statement by a careful examination of the specimens, and has ascertained that there are two very distinct species of this remarkable type of fishes.

Tales of "Wonderful Fish" are common in the works of the old naturalists, whence they are quoted from generation to generation. Sir John Richardson has lately demolished one queer fish, which was as certain to reappear whenever opportunity offered, as the elephant pricked with the tailor's needle does in books of stories of the animal world. We allude to that monstrous myth, the great Manheim Pike, with a collar round his neck, put into a lake by the Emperor Frederick II. in the year 1230; and taken out in the 276th year of his age, the 17th foot of his length, and the 350th pound of his weight. M. Valenciennes, a naturalist of repute, has entered into a critical history of this monster, and has found him to be apocryphal. The creature was, at any rate, taken in several places at once, the legends written on his brass collar do not agree, and his alleged skeleton has been found to be made up of various bones of various fishes; while the vertebræ are, unfortunately, so many, that Professor Owen would order him out of Court in an instant as a rank impostor. Probably some specimen of the *Mecho*, the monstrous fish of the Danube—which has even now been scarcely described, and which has only recently been identified as one of the salmon tribe—having been called a pike, may be at the bottom of the legend of the great Manheim fish. But Sir John Richardson produces another big pike, killed by an intrepid " angler seventy years of age, with a single rod and bait "—an observation which leads to the inquiry of the possibility of catching a single fish with more than one rod and bait—" that weighed seventy-eight pounds." This is stated to have happened in the county of Clare; the angler's name was O'Flanagan.

Here is another wonderful story:—The Bohemians have a proverb—"Every fish has another for prey:" that named the Wels has them all. This is the largest fresh-water fish found in the rivers of Europe, except the sturgeon; it often reaches five or six feet in length. It destroys many aquatic birds, and we are assured that it does not spare the human species. On the 3d of July, 1700, a peasant took one near Thorn, that had an infant entire in its stomach! They tell in Hungary of children and young girls being devoured on going to draw water; and they even relate that, on the frontiers of Turkey, a poor fisherman took one that had in its stomach the body of a woman, her purse full of gold, and a *ring!* The fish is even reported to have been taken sixteen feet long. The old stories of rings found in the stomachs of fishes will be remembered; as well as here and there a *book* found in the stomach of a fish!

The Sun-fish is exceedingly rare. A large specimen was captured off Start Point in 1864. Attention was first drawn to a huge dark object on the water. On a boat being sent out, it was soon discovered to be the back fin of a very large fish, apparently asleep. A very exciting chase commenced, extending over an hour, the crew meanwhile battling with harpoons, boat-hooks, &c.; the fish trying several times to upset the boat by getting his back under it. At length a line was thrown over its head, and the fish, being weakened by the struggle, was towed alongside the yacht, hoisted on board, and slaughtered. Yarrell, in his work on British Fishes, states the largest Sun-fish to be about 3 cwt., but the above specimen weighed nearly 6 cwt. Sun-fish are found occasionally in the tropical seas of large dimensions, but those found in the Channel seldom if ever exceed from 1 cwt. to 2 cwt. The peculiarities in regard to this fish are, that it has no bones, but the whole of the formation is of cartilage, which can easily be cut with a knife. The skin is cartilage of about an inch and a-half thick, under which there is no backbone or ribs. This specimen was of extraordinary dimensions—5 ft. 10 in. in length, and 7 ft. from the tip of the dorsal to the point of the anal fin.

The "Courrier de Sagon" brings, as a contribution to Natural History, the not very credible-sounding description of a fish called "Ca-oug" in the Anamite tongue, which is said to have saved the lives already of several Anamites; for which reason the King of Anam has invested it with the name of "Nam hai dui bnong gnan" (Great General of the South Sea). This fish is said to swim round ships near the coast, and, when it sees a man in the water, to seize him with his mouth, and to carry him ashore. A skeleton of this singular inhabitant of the deep is to be seen at Wung-tau, near Cape St. James. It is reported to be thirty-five feet in length, to have tusks "almost like an elephant," very large eyes, a black and smooth skin, a tail like a lobster, and two "wings" on its back.*

The Grouper must be a voracious fish, for we read of a specimen being caught off the coast of Queensland, which is thus described:—"It was 7 ft. long, 6 ft. in circumference at its thickest part, and its head weighed 80 lb. When opened, there were found in its stomach two broken bottles, a quart pot, a preserved milk tin, seven medium-sized crabs; a piece of earthenware, triangular in shape, and three inches in length, incrusted with oyster shells, a sheep's head, some mutton and beef bones, and some loose oyster shells. The spine of a skate was imbedded in the Grouper's liver."

The Double-fish, here represented, is a pair of Cat-fish, which were taken alive in a shrimp-net, at the mouth of Cape Fear River, near Fort Johnston, North Carolina, in 1833, and presented to Professor Silliman. One of them is three and a-half, and the

* "Athenæum."

other two and a-half inches long, including the tail —the smallest emaciated, and of sickly appearance. They are connected in the manner of the Siamese Twins, by the skin at the breast, which is marked by a dark streak at the line of union. The texture and colour otherwise of this skin is the same as that of the belly. The mouth, viscera, &c., were entire and perfect in each fish; but, on withdrawing the entrails, through an incision made on one side of the abdomen, the connecting integument was found to be hollow. A flexible probe was passed through from one to the other, with the tender and soft end of a spear of grass, drawn from a green plant. But there was no appearance of the entrails of one having come in contact with those of the other, for the integument was less than one-tenth of an inch in its whole thickness; in length, from the body or trunk of one fish to the other, it was three-tenths; and in the water, when the largest fish was in its natural position, the small one could, by the length and pliancy of this skin, swim in nearly the same position. When these fish came into existence it is probable they were of almost equal size and strength, but one "born to better fortune," or exercising more ingenuity and industry than the other, gained a trifling ascendency, which he improved to increase the disparity, and, by pushing his extended mouth in advance of the other, seized the choicest and most of the food for himself.

From the northern parts of British America we have received extraordinary contributions to our fish collections. One of these is the Square-browed Malthe, obtained in one of the land expeditions under the command of Captain Sir John Franklin, R.N. It was taken on the Labrador coast, and then belonged to a species hitherto undescribed. Its intestines were filled with small crabs and univalve shells. The extreme length of the fish is 7 inches 11 lines. The upper surface is greyish white, with brown blotches, and the fins are whitish. The head is much depressed and greatly widened; the eyes far forward; the snout projecting like a small horn. Most of the fish of this family can live long out of water, in consequence of the smallness of their gill-openings; indeed, those of one of the genera are able, even in warm countries, to pass two or three days in creeping over the land. All the family conceal themselves in the mud or sand, and lie in wait to take their prey by surprise. The accompanying engraving is from the very able work of Dr. Richardson, F.R.S., published by the munificence of Government.

Gold Fish (of the Carp family) have been made to distinguish a particular sound made by those from whom they receive their food; they recognise their footsteps at a distance, and come at their call. Captain Brown says Gold Fish, when kept in ponds, are "frequently taught to rise to the surface of the water at the sound of a bell to be fed;" and Mr. Jesse was assured that Gold Fish evince much pleasure on being whistled to. Hakewill, in his "Apology for God's Power and Providence,"

SQUARE-BROWED MALTHE AND DOUBLE FISH.

cites Pliny to show that a certain emperor had ponds containing fish, which, when called by their respective *names* that were bestowed upon them, came to the spot whence the voice proceeded. Bernier, in his " History of Hindustan," states a like circumstance of the fish belonging to the Great Mogul. The old poet, Martial, also mentions fish coming at the call, as will be seen by the following translation from one of his epigrams :—

" Angler ! could'st thou be guiltless ? Then forbear :
For these are sacred fishes that swim here ;
Who know their Sovereign, and will lick his hand,
Than which none's greater in the world's command ;
Nay, more ; they've names, and when they called are,
Do to their several owners' call repair."

Who, after reading so many instances, can doubt that fish hear ?

It has been found that the water from steam-engines, which is thrown into dams or ponds for the purpose of being cooled, conduces much to the nutriment of Gold Fish. In these dams, the average temperature of which is about eighty degrees, it is common to keep Gold Fish ; in which situation they multiply much more rapidly than in ponds of lower temperature exposed to variations of the climate. Three pair of fish were put into one of these dams, where they increased so rapidly that at the end of three years their progeny, which was accidentally poisoned by verdigris mixed with the refuse tallow from the engine, were taken out by wheel-barrow-fuls. Gold Fish are by no means useless inhabitants of these dams, as they consume the refuse grease which would otherwise impede the cooling of the water by accumulating on its surface. It is not improbable that this unusual supply of aliment may co-operate with increase of temperature in promoting the fecundity of the fishes.

Most of our readers have heard of the fish popularly known as the Miller's Thumb, the origin of the name of which Mr. Yarrell has thus explained:—"It is well known that all the science and tact of a miller is directed so to regulate the machinery of his mill that the meal produced shall be of the most valuable description that the operation of grinding will permit, when performed under the most advantageous circumstances. His ear is constantly directed to the note made by the running stone in its circular course over the bedstone, the exact parallelism of their two surfaces, indicated by a particular sound, being a matter of the first consequence; and his hand is constantly placed under the meal-spout to ascertain, by actual contact, the character and quality of the meal produced, which he does by a particular movement of his thumb in spreading the sample over his fingers. By this incessant action of the miller's thumb, a peculiarity in its shape is produced, which is said to resemble exactly the shape of the *river bull-head*, a fish constantly found in the mill-stream, and which has obtained for it the name of the Miller's Thumb."

M. Coste has constructed a kind of marine observatory at Concarneau (Finisterre) for the purpose of studying the habits and instincts of various Sea-fish. A terrace has been formed on the top of a house on the quay, with reservoirs arranged like a flight of steps. The sea-water is pumped up to the topmost reservoir, and thence flows down slowly, after the manner of a rivulet. The length is divided into 95 cells by wire net partitions, which, allowing free passage to the water, yet prevent the different species of fish from mingling together. By this ingenious contrivance each kind lives separate, enjoying its peculiar food and habits, unconscious of its state of captivity. Some species, such as the Mullet, the Stickleback, &c., grow perfectly tame, will follow the hand that offers them food, and will even allow themselves to be taken out of the water. The Goby and Bull-head are less familiar. The Turbot, which looks so unintelligent, will, nevertheless, take food from the hand; it changes colour when irritated, the spots with which it is covered growing pale or dark, according to the emotions excited in it. But the most curious circumstance concerning it is, that it swallows fish of a much larger size than would appear compatible with the apparent smallness of its mouth. Thus, a young Turbot, not more than ten inches in length, has been seen to swallow Pilchards of the largest size. The Pipe-Fish has two peculiarities. These fish form groups, entwining their tails together, and remaining immoveable in a vertical position, with their heads upwards. When food is offered them, they perform a curious evolution—they turn round on their backs to receive it. This is owing to the peculiar position of the mouth, which is placed under a kind of beak, and perpendicular to its axis.

The crustaceous tribes have also furnished much matter of observation. The Prawn and Crab, for instance, exercises the virtue of conjugal fidelity to the highest degree; for the male takes hold of his mate, and never lets her go; he swims with her, crawls about with her, and if she is forcibly taken away from him, he seizes hold of her again. The metamorphoses to which various crustaceous tribes are subject have also been studied with much attention.*

Much as the nature and habits of fish have been studied of late years, the economy of some is to this day involved in obscurity. The Herring is one of these fishes. The Swedish Herring Fisheries were, at one time, the largest in Europe, but at present, during the temporary disappearance of the fish, they have dwindled away. The causes which influence the movements of the Herring—one of the most capricious of fish—are a puzzle which naturalists have as yet failed to solve. They are not migratory, as was at one time believed—that is, they seldom wander far from the place where they were bred; but they are influenced by certain hidden and unexplained causes at one time to remain for years in the deep sea, and at another to come close in to land in enormous numbers. During the first half of the sixteenth century, Herrings entirely deserted the

* See "The Tree-climbing Crab," pp. 282-302.

Swedish coasts. In 1556 they reappeared, and remained for thirty-one years in the shallow waters. Throughout this period they were taken in incalculable numbers; "thousands of ships came annually from Denmark, Germany, Friesland, Holland, England, and France, to purchase the fish, of which sufficient were always found for them to carry away to their own or other countries. . . . From the small town of Marstrand alone some two million four hundred thousand bushels were yearly exported." In 1587 the Herrings disappeared, and remained absent for seventy-three years, till 1660. In 1727 they returned, and again in 1747, remaining till 1808, and during this last period the fisheries were prosecuted with extraordinary zeal, industry, and success. The Government gave every encouragement to settlers, and it was computed that during some years as many as fifty thousand strangers took part in them. In 1808 the Herrings once more disappeared, and have never returned since. The cause must still be considered as quite unknown; but we may fairly assume, according to historical precedents, that after a certain period of absence, the Herrings will again return.*

Aristotle, in his "History of Animals," makes some extremely curious observations on Fish and Cetaceous Animals, as might be expected from the variety of these animals in the Grecian seas. In Spratt and Forbes's "Travels in Syria" the account of the habits and structure of the Cuttle-fish in

* "Saturday Review."

Aristotle's work is ranked amongst the most admirable natural history essays ever written. It is, moreover, remarkable for its anticipation.

Dr. Osborne, in 1840, read to the Royal Society a short analysis of this work, in which he showed that Aristotle anticipated Dr. Jenner's researches respecting the cuckoo; as also some discoveries respecting the incubated egg, which were published as new in the above year. Aristotle describes the economy of bees as we have it at present; but mistakes the sex of the queen. The various organs are described as modified throughout the different classes of animals (beginning with man) in nearly the same order as that afterwards adopted by Cuvier.

The chief value of this body of knowledge, which has been buried for above 2,000 years, is, that it is a collection of facts observed under peculiar advantages, such as never since occurred, and that *it is at the present day to be consulted for new discoveries.*

According to Pliny, for the above work some thousands of men were placed at Aristotle's disposal throughout Greece and Asia, comprising persons connected with hunting and fishing, or who had the care of cattle, fish-ponds, and apiaries, in order that he might obtain information from all quarters, *ne quid usquam gentium ignoretur ab eo.* According to Athenæus, Aristotle received from the prince, on account of the expenses of the work, 800 talents, or upwards of 79,000*l.*

FISH IN BRITISH COLOMBIA.

IN this bitterly cold country, where the snow lies deep six months out of the twelve, the natives subsist principally on fish, of which there is an extraordinary abundance generally, and of salmon particularly. Salmon swarm in such numbers that the rivers cannot hold them. In June and July every rivulet, no matter how shallow, is so crammed with salmon that, from sheer want of room, they push one another high and dry upon the pebbles; and Mr. Lord * tells us that each salmon, with its head up, struggles, fights, and scuffles for precedence. With one's hands only, or more easily by employing a gaff or a crook-stick, tons of salmon have been procured by the simple process of hooking them out. Once started on their journey, the salmon never turn back. As fast as those in front die, fresh arrivals crowd on to take their places,

* "The Naturalist in Vancouver Island and British Columbia." By John Keast Lord, F.Z.S., Naturalist to the British North American Boundary Commission.

and share their fate. "It is a strange and novel sight to see three moving lines of fish—the dead and dying in the eddies and slack water along the bank, the living breasting the current in the centre, blindly pressing on to perish like their kindred." For two months this great *salmon army* proceeds on its way up stream, furnishing a supply of food without which the Indians must perish miserably. The winters are too severe for them to venture out in search of food, even if there was any to be obtained. From being destitute of salt, they are unable to cure meat in the summer for winter provisions, and hence for six months in the year they depend upon salmon, which they preserve by drying in the sun.

But the Indian has another source of provision for the winter, fully as important as the salmon. The Candle-fish supplies him at once with light, butter, and oil.* When dried, and perforated with a rush, or strip of cypress-bark, it can be lighted, and burns steadily until consumed. Strung up, and hung for a time in the smoke of a wood fire, it is preserved as a fatty morsel to warm him when pinched with cold; and, by heat and pressure, it is easily converted into liquid oil, and drunk with avidity. That nothing may be wanting, the hollow stalk of the sea-wrack, which at the root is expanded into a complete flask, makes an admirable bottle; and so, when the Indian buries himself for long dreary months in his winter quarters, neither his

* The Petrel is similarly used in the Faroe Islands. (See *ante*, p. 234.) It may, therefore, be called the Candle Bird.

larder nor his cellar are empty, and he has a lamp to lighten the darkness. The steamers have, however, frightened away the Candle-fish and the Indian from their old haunts, and they have both retreated to the north of the Colombia River.

Amongst the other inhabitants of the salt and fresh waters of these regions are the Halibut and the Sturgeon, both of which attain to an immense size. The bays and inlets along the coast abound with marine wonders. There feasts and fattens the Clam, a bivalve so gigantic that no oyster-knife can force an entrance, and only when his shell is almost red-hot will he be at last constrained to open his dwelling.

And there lies in wait the awful Octopus, a monster of insatiable voracity, of untameable ferocity, and of consummate craft; of sleepless vigilance, shrouded amidst the forest of sea-weed, and from the touch of whose terrible arms no living thing escapes. It attains to an enormous size in those seas, the arms being sometimes five feet in length, and as thick at the base as a man's wrist. No bather would have a chance if he once got within the grasp of such a monster, nor could a canoe resist the strength of its pull; but the Indian, who devours the Octopus with great relish, has all the cunning created by necessity, and takes care that none of the eight sucker-dotted arms ever gain a hold on his frail bark.

Professor Owen has figured a species of Octopus, the Eight-armed Cuttle of the European seas, representing it in the act of creeping on shore, its body being carried vertically in the reverse position, with its head downwards, and its back being turned towards the spectator, upon whom it is supposed to be advancing. This animal is said to be luminous in the dark. Linnæus quotes Bartholinus for the statement that one gave so much light that when the candle was taken away, it illuminated the room.

The Sturgeon is one of the finest fishes of the country, and Mr. Lord's account of the Indian mode of taking them is a very graphic picture of this river sport.

"The spearman stands in the bow, armed with a most formidable spear. The handle, from seventy to eighty feet long, is made of white pine-wood ; fitted on the spear-haft is a barbed point, in shape very much like a shuttlecock, supposing each feather represented by a piece of bone, thickly barbed, and very sharp at the end. This is so contrived that it can be easily detached from the long handle by a sharp, dexterous jerk. To this barbed contrivance a long line is made fast, which is carefully coiled away close to the spearman, like a harpoon-line in a whale-boat. The four canoes, alike equipped, are paddled into the centre of the stream, and side by side drift slowly down with the current, each spearman carefully feeling along the bottom with his spear, constant practice having taught the crafty savages to know a Sturgeon's back when the spear comes in contact with it. The spear-head touches the drowsy fish ; a sharp plunge, and the redskin sends the notched points through armour and cartilage, deep into the leather-like muscles. A skilful jerk frees the long handle from the barbed end, which remains inextricably fixed in the fish ; the handle is thrown aside, the line seized, and the struggle begins. The first impulse is to resist this objectionable intrusion, so the angry Sturgeon comes up to see what it all means. This curiosity is generally repaid by having a second spear sent crashing into him. He then takes a header, seeking safety in flight, and the real excitement commences. With might and main the bowman plies the paddle, and the spearman pays out the line, the canoe flying through the water. The slightest tangle, the least hitch, and over it goes; it becomes, in fact, a sheer trial of paddle *versus* fin. Twist and turn as the Sturgeon may, all the canoes are with him. He flings himself out of the water, dashes through it, under it, and skims along the surface; but all is in vain, the canoes and their dusky oarsmen follow all his efforts to escape, as a cat follows a mouse. Gradually the Sturgeon grows sulky and tired, obstinately floating on the surface. The savage knows he is not vanquished, but only biding a chance for revenge; so he shortens up the line, and gathers quietly on him to get another spear in. It is done,—and down viciously dives the Sturgeon; but pain and weariness begin to tell, the struggles grow weaker and weaker as life ebbs slowly away, until the mighty armour-plated monarch of the river yields himself a captive to the dusky native in his frail canoe."

There is a very rare Spoonbill Sturgeon found in the western waters of North America: its popular name is Paddle-fish. One, five feet in length, weighed forty pounds; the nose, resembling a spatula, was thirteen inches in length. It was of a light slate colour, spotted with black; belly white; skin smooth, like an eel; the flesh compact and firm, and hard when boiled—not very enticing to the epicure. The jaws are without teeth, but the fauces are lined with several tissues of the most beautiful network, evidently for the purpose of collecting its food from the water by straining, or passing it through these membranes in the same manner as practised by the spermaceti whale. Near the top of the head are two small holes, through which it is possible the Sturgeon may discharge water in the manner practised by cetaceous animals. It is conjectured that the long "Spoonbill" nose of this fish is for digging up or moving the soft mud in the bottom of the river, and when the water is fully saturated, draw it through the filamentory strainers in search of food.

Sturgeons resemble sharks in their general form, but their bodies are defended by bony shields, disposed in longitudinal rows; and their head is also well curiassed externally. The Sturgeons of North America are of little benefit to the natives. A few speared in the summer-time suffice for the temporary support of some Indian hordes; but none are preserved for winter use, and the roe and sounds are utterly wasted.

The northern limit of the Sturgeon in America is probably between the 55th and 56th parallels of latitude. Dr. Richardson did not meet with any account of its existence to the north of Stewart's Lake, on the west side of the Rocky Mountains; and on the east side it does not go higher than the Saskatchewan and its tributaries. It is not found in Churchill River, nor in any of the branches of the Mackenzie or other streams that fall into the Arctic Seas—a remarkable circumstance when we consider that some species swarm in the Asiatic rivers which flow into the Icy Sea. Sturgeons occur in all the great lakes communicating with the St. Lawrence, and also along the whole Atlantic coast of the United States down to Florida. Peculiar species inhabit the Mississippi; it is, therefore, probable that the range of the genus extends to the Gulf of Mexico.

The great rapid which forms the discharge of the Saskatchewan into Lake Winnipeg appears quite alive with these fish in the month of June; and some families of the natives resort thither at that time to spear them with a harpoon, or grapple them with a strong hook tied to a pole. Notwithstanding the great muscular power of the Sturgeon, it is timid; and Dr. Richardson saw one so frightened at the paddling of a canoe, that it ran its nose into a muddy bank, and was taken by a *voyageur*, who leaped upon its back.

In Colombia River, a small species of Sturgeon attains eleven feet in length, and a weight of six hundred pounds.* It is caught as high up as Fort Colville, notwithstanding the numerous intervening cataracts and rapids which seem to be insuperable barriers to a fish so sluggish in its movements.

The Sturgeon is styled a Royal Fish in England, because, by a statute of Edward II. it is enacted, " the King shall have Sturgeon taken in the sea, or elsewhere, within the realm."

---

* Dr. Richardson. The *Huro* is reported by Pallas to attain a weight of nearly three thousand pounds, and a length exceeding thirty feet.

THE TREE-CLIMBING CRAB.

THE transition from the ordinary mode of the locomotion of fishes by swimming to that of climbing has been ably illustrated by the Rev. Dr. Buckland, who showed, in a communication to the Ashmolean Society, in 1843, that the fins in certain genera perform the functions of feet and wings. Thus, "fishing-frogs" have the fins converted into feet, or paddles, by means of which they have the power of crawling or hopping on sand and mud; and another species can live three days out of the water, and walk upon dry land. The climbing perch of the Indian rivers is known to live a long time in the air, and to climb up the stems of palm-trees in pursuit of flies, by means of spinous projections on its gill-covers. Fishes of the *silurus* family have a bony enlargement of the first ray of the pectoral fin, which is also armed with spines; and this is not only an offensive and defensive weapon, but enables the fish to walk along the bottom of the fresh waters

THE TREE-CLIMBING CRAB.

which it inhabits. The flying-fishes are notorious examples of the conversion of fins into an organ of movement in the air. M. Deslongchamps has published, in the " Transactions of the Linnæan Society of Normandy," 1842, a curious account of the movements of the gurnard at the bottom of the sea. In 1839, he observed these movements in one of the artificial fishing-ponds, or fishing-traps, surrounded by nets, on the shore of Normandy. He saw a score of gurnards closing their fins against their sides, like the wing of a fly in repose, and without any movement of their tails, walking along the bottom by means of six free rays, three on each pectoral fin, which they placed successively on the ground. They moved rapidly forwards, backwards, to the right and left, groping in all directions with these rays, as if in search of small crabs. Their great heads and bodies seemed to throw hardly any weight on the slender rays, or feet, being suspended in water, and having their weight further diminished by their swimming-bladder. During these movements the gurnards resembled insects moving along the sand. When M. Deslongchamps moved in the water, the fish swam away rapidly to the extremity of the pond; when he stood still, they resumed their ambulatory movement, and came between his legs. On dissection, we find these three anterior rays of the pectoral fins to be supported each with strong muscular apparatus to direct their movements, apart from the muscles that are connected with the smaller rays of the pectoral fin.

T

Dr. Buckland states that Miss Potts, of Chester, had sent to him a flagstone from a coalshaft at Mostyn, bearing impressions which he supposed to be the trackway of some fish crawling along the bottom by means of the anterior rays of its pectoral fins. There were no indications of feet, but only scratches, symmetrically disposed on each side of a space that may have been covered by the body of the fish whilst making progress, by pressing its fin-bones on the bottom. As yet, no footsteps of reptiles, or of any animals more highly organized than fishes, have been found in strata older than those which belong to the new red sandstone. The abundant remains of fossil fishes, armed with strong bony spines, and of other fishes allied to the gurnard, in strata of the carboniferous and old red sandstone series, would lead us to expect the frequent occurrence of impressions made by their locomotive organs on the bottoms of the ancient waters in which they lived. Dr. Buckland proposed to designate these petrified traces or trackways of ancient fishes by the term of fish-tracks.

Crabs and Lobsters are strange creatures: strange in their configurations; strange in the transmutations which they exhibit from the egg to maturity; strange in the process they undergo of casting off, not only their shell, but the covering of their eyes, of their long horns, and even the lining of their tooth-furnished stomach; strange, also, are they in their manners and habits. Many a reader, in wandering along the sea-shore, may have disturbed little colonies of Crabs quietly nestling in fancied security amidst banks of slimy sea-weed; and in the nooks and recesses of the coast, the shallows, and strips of land left dry at ebb-tide, may be seen numbers of little, or perchance large, Crabs, some concealed in snug lurking-places, others tripping, with a quick *side-long* movement, over the beach, alarmed by the advance of an unwelcome intruder. Some are exclusively tenants of the water, have feet formed like paddles for swimming, and never venture on land; others seem to love the air and sunshine, and enjoy an excursion, not without hopes of finding an acceptable repast, over the oozy sands; some, equally fond of the shore and shallow water, appropriate to themselves the shells of periwinkles, whelks, &c., and there live in a sort of castle, which they drag about with them on their excursions, changing it for a larger as they increase in measure of growth. They vary in size from microscopic animalcules to the gigantic King Crab :* to the former, the luminosity of the ocean, or of the foam

---

* This Crab has an elongated spine-like tail, the use of which was long misunderstood. Dr. J. Gray was shown at the Liverpool Museum some living King Crabs, and the use they made of the tail-like appendages. When turned over on their backs, he saw them bend down the tail until they could reach some point of resistance, and then employ it to elevate the body, and regain their normal position. Dr. Gray states that they never have been seen to use this tail for the purpose which has been often assigned to it—that is, for leaping from place to place by bending it under the body, like the toy called a "spring-jack," or "leaping frog."

before the prows of vessels, is, to a great extent, attributable, each minute creature glowing with phosphoric light.

The Bernhard Crab has been proved to have the power of dissolving shells, it not being unusual to find the long fusiform shells which are inhabited by these animals with the inner lip, and the greater part of the pillar on the inside of the mouth, destroyed, so as to render the aperture much larger than usual. Dr. Gray is quite convinced that these Crabs have the above power, some to a much greater degree than others.

Certain Crabs, especially in the West Indies, are almost exclusively terrestrial, visiting the sea only at given periods, for the deposition of their eggs. These Crabs carry in their gill-chambers sufficient water for the purpose of respiration; they live in burrows, and traverse considerable tracts of land in the performance of their migratory journeys. Of these, some, as the Violet Crab, are exquisite delicacies.

Of a great Crab migration we find these details in the "Jamaica Royal Gazette:"—In 1811 there was a very extraordinary production of Black Crabs in the eastern part of Jamaica. In June or July the whole district of Manchidneed was covered with countless numbers, swarming from the sea to the mountains. Of this the writer was an eye-witness. On ascending Over Hill from the vale of Plantain Garden River, the road appeared of a reddish colour, as if strewed with brick-dust. It was owing to myriads of young Black Crabs, about the size of the nail of a man's finger, moving at a pretty quick pace, direct for the mountains. "I rode along the coast," says the writer, "a distance of about fifteen miles, and found it nearly the same the whole way. Returning the following day, I found the road still covered with them, the same as the day before. How have they been produced, and where do they come from? were questions everybody asked, and nobody could answer. It is well known that Crabs deposit their eggs once a year, in May; but, except on this occasion, though living on the coast, I had never seen above a dozen young Crabs together; and here were myriads. No unusual number of old Crabs had been observed in that season; and it is worthy of note, that they were moving from a rockbound coast of inaccessible cliffs, the abode of seabirds, and exposed to the constant influence of the trade winds. No person, as far as I know, ever saw the like, except on that occasion; and I have understood that since 1811 Black Crabs have been more abundant further in to the interior of the island than they were ever known before."

Cuvier describes the Burrowing Crab as displaying wonderful instinct:—"The animal closes the entrance of its burrow, which is situated near the margin of the sea, or in marshy grounds, with its largest claw. These burrows are cylindrical, oblique, very deep, and very close to each other; but generally each burrow is the exclusive habitation of a single individual. The habit which these crabs have of holding their large claw elevated in advance of the body, as if making a sign of beckoning to some one, has obtained for them the name of Calling Crabs. There is a species observed by Mr. Bosc in South Carolina, which passes the three months of the winter in its retreat without once quitting it, and which never goes to the sea except at the epoch of egg-laying." The same observations apply to the Chevalier Crabs (so called from the celerity with which they traverse the ground). These are found in Africa, and along the borders of the Mediterranean.

Some Crabs, truly aquatic, as the Vaulted Crab of the Moluccas, have the power of drawing back their limbs, and concealing them in a furrow, which they closely fit; and thus, in imitation of a tortoise, which retracts its feet and head within its shell, they secure themselves, when alarmed. Other aquatic species have their limbs adapted for clinging to weeds and other marine objects. Of these some have the two or four hind pairs of limbs so placed as to appear to spring from the back; they terminate in a sharp hook, by means of which the Crab attaches itself to the valves of shells, fragments of coral, &c., which it draws over its body, and thus lurks in concealment. Allied, in some respects, to the Hermit or Soldier Crabs, which tenant empty shells, is one which, from its manners and habits, is one of the most extraordinary of its race. The Hermit Crabs are voracious, and feed on animal substances. The Hermit, or Bernhard Crab, is so called from its habit of taking up its solitary residence in deserted shells, thus seeking a protection for its tail, which is long and naked. It is found in shells of different dimensions, and from time to time leaves its abode, as it feels a necessity, for a more commodious dwelling. It is said to present, on such occasions, an amusing instinct as it inserts the tail successively into several empty shells until one is found to fit. We learn from Professor Bell, however, that it does not always wait until the home is vacant, but occasionally rejects the rightful occupant with some violence. On the contrary, the Crab, or rather Lobster-Crab (for it takes an intermediate place between them), is more delicate in its appetite, and feeds upon fruits, to obtain which it is said to climb up certain trees, at the feet of which it makes a burrow. This species is the Purse Crab, or Robber Crab, of Amboyna and other islands in the South Pacific Ocean.

"According to popular belief among the Indians," says Cuvier, "the Robber Crab feeds on the nuts of the cocoa-tree, and it makes its excursions during the night; its places of retreat are fissures in the rocks, or holes in the ground." The accounts of the early writers and travellers, as well as of the natives, were disbelieved; but their truth has since been abundantly confirmed. MM. Quoy and Guimard assure us that several Robber Crabs were fed by them for many months on cocoa-nuts alone; and a specimen of this Crab was submitted to the Zoological Society, with additional information from

Mr. Cuming, in whose fine collection from the islands of the South Pacific several specimens were preserved. Mr. Cuming states these Crabs to be found in great numbers in Lord Hood's Island, in the Pacific. He there frequently met with them on the road. On being disturbed, the Crabs instantly assumed a defensive attitude, making a loud snapping with their powerful claws, or pincers, which continued as they retreated backwards. They climb a species of palm to gather a small kind of cocoa-nut that grows thereon. They live at the roots of trees, and not in the holes of rocks; and they form a favourite food among the natives. Such is the substance of Mr. Cuming's account. Mr. Darwin, in his "Researches in Geology and Natural History," saw several of these Crabs in the Keeling Islands, or Cocos Islands, in the Indian Ocean, about 600 miles distant from the coast of Sumatra. In these islands, of coral formation, the cocoa-nut tree is so abundant as to appear, at first glance, to compose the whole wood of the islands.

Here the great Purse Crab is abundant. Mr. Darwin describes it as a Crab which lives on the cocoa-nut, is common on all parts of the dry land, and grows to a monstrous size. This Crab has its front pair of legs terminated by very strong and heavy pincers, and the last pair by others which are narrow and weak. It would at first be thought quite impossible for a Crab to open a strong cocoa-nut, covered with the husk; but Mr. Liesk assures me that he has repeatedly seen the operation effected. The Crab begins by tearing away the husk, fibre by fibre, and always from that end under which the three eye-holes are situated. When this is completed the Crab commences hammering with its heavy claws on one of these eye-holes till an opening is made. Then, turning its body, by the aid of its posterior and narrow pair of pincers, it extracts the white albuminous substance. I think this as curious a case as I ever heard of, and likewise of adaptation in structure between two objects apparently so remote from each other in the scheme of nature as a Crab and a cocoa-nut tree. The Crab is diurnal in its habits; but it is said to pay every night a visit to the sea for the purpose of moistening its gills. These gills are very peculiar, and scarcely fill up more than a tenth of the chamber in which they are placed: it doubtless acts as a reservoir for water, to serve the Crab in its passage over the dry and heated land. The young are hatched and live for some time on the coast; at this period of existence we cannot suppose that cocoa-nuts form any part of their diet; most probably soft saccharine grasses, fruits, and certain animal matters, serve as their food until they attain a certain size and strength.

The adult Crabs, Mr. Darwin tells us, inhabit deep burrows, which they excavate beneath the roots of trees; and here they accumulate great quantities of the picked fibres of the cocoa-nut husk, on which they rest as on a bed. The Malays sometimes take advantage of the labours of the Crab by collecting the coarse fibrous substance, and using it as junk.

These Crabs are very good to eat; moreover, under the tail of the larger ones there is a great mass of fat, which, when melted, yields as much as a quart bottleful of limpid oil.

The Crab's means of obtaining the cocoa-nuts have, however, been much disputed. It is stated by some authors to crawl up the trees for the purpose of stealing the nuts. This is doubted; though in the kind of palm to which Mr. Cuming refers as being ascended by this Crab, the task would be much easier. Now, Mr. Darwin states, that in the Keeling Islands the Crab lives only on the nuts which fall to the ground. It may thus appear that Mr. Cuming's and Mr. Darwin's respective accounts of the *non-climbing* of this Crab on the one side, and its *actually climbing trees* on the other, are contradictory. The height of the stem of the cocoa-nut tree, its circumference, and comparative external smoothness, would prove insurmountable, or at least very serious obstacles, to the most greedy Crab, however large and strong it might be. But these difficulties are by no means so formidable in the tree specified by Mr. Cuming: this is arborescent, or bushy, with long, thin, rigid, sword-shaped leaves, resembling those of the pineapple, usually arranged spirally, so that they are commonly called Screw Pines. They are of the genus *Pandanus*, a word derived from the Malay *Pandang*. The ascent of these arborescent plants, having the stem furnished with a rigging of cord-like roots, and bearing a multitude of firm, long, and spirally-arranged leaves, would be by no means a work of difficulty, as would necessarily be that of the tall feathery-topped cocoa-tree, destitute of all available points of aid or support. Hence the contradiction in the two accounts referred to is seeming, and not real, and the two statements are reconciled.

To sum up, Mr. Cuming fully testifies to the Crab climbing the Screw Pines; and he has told Professor Owen that he has actually seen the Crab climbing the cocoa-nut tree. The Crab has been kept on cocoa-nuts for months; and is universally reported by the natives to climb the trees at night.

We may here, too, observe, that fine specimens of the Climbing Crab are to be seen in the British Museum. Here, too, arranged in cases, are Spider Crabs; Crabs with oysters growing on their backs, thus showing that Crabs do not shed their shells every year, or that the oyster increases very rapidly in bulk; Oval-bodied Crabs; and Fin-footed or Swimming Crabs. Here are also Telescope, or Long-eyed Crabs, and Land Crabs, found in India 4,000 feet above the sea-level; another of similar habits in the plains of the Deccan, that may be seen swarming in the fields, some cutting and nipping the green rice-stalks, and others waddling off backwards with sheaves bigger than themselves. To these may be added Square-bodied Crabs, Crested Crabs; Porcelain Crabs, with delicate, china-like shells; and Death's-head Crabs, which usually form cases for themselves from pieces of sponge and shells.

Certain species of Crabs are remarkably tenacious of life, and have been known to live for weeks buried, and without food. It is in the Crab tribe that the fact of the metamorphosis of *crustacea* has been most distinctly perceived; a small, peculiar crustacean animal, that had long passed for a distinct species, under the name of *Zoea*, having at length been identified with the young of the common Crab before it had attained its full development.

That among the Crab tribes a tree-climbing species is to be found is certainly curious, but it is not without a parallel among fishes. Many of the latter leave the water, some even for a long time, and perform overland journeys, aided in their progress by the structure of their fins. In these fishes the gills and gill-chambers are constructed for the retention of water for a considerable time, so as to suffice for the necessary degree of respiration. In our country, we may mention the eel, which often voluntarily quits the river or lake, and wanders during the night over the adjacent meadows, probably in quest of dew-worms. But the marshes of India and China present us with fishes much more decidedly terrestrial, and some of which were known to the ancients. Among these are several fishes of a snake-like form : they have an elongated, cylindrical body, and creep on land to great distances from their native waters. The boatmen of India often keep these fishes for a long time out of water, for the sake of diverting themselves and others by their terrestrial movements, and children may often be seen enjoying this sport.

Of these land-haunting fishes, the most remarkable

is the tree-climber, so called in Tranquebar. This fish inhabits India, the Indian islands, and various parts of China, as Chusan, &c., living in marshes, and feeding on aquatic insects, worms, &c. According to Daldorf, a Danish gentleman, who, in 1797, communicated an account of the habits of this fish to the Linnæan Society, it *mounts up* the bushes or low palms to some elevation. This gentleman states that he had himself observed it in the act of ascending palm-trees near the marshes, and had taken it at a height of no less than five feet, measured from the level of the adjacent water. It effects its ascent by means of its pectoral and under fins, aided by the action of the tail and the spines which border the gill covers. It is by the same agency that it traverses the land. The statement of M. Daldorf is corroborated by M. John, also a Danish observer, to whom we are indebted for the knowledge of its name in Tranquebar, which alludes to its arboreal proceedings.

It is true that many other naturalists who have observed the habits of this fish in its native regions, while they concur in describing its terrestrial journeys, and its living for a long time out of water, either omit to mention, or mention with doubt, its reputed attempts at *tree-climbing*.

The habits and instincts of certain Crawfishes are very extraordinary. Thus, the *Astaci* are migratory, and in their travels are capable of doing much damage to dams and embankments. On the Little Genesee River they have, within a few years, compelled the owner of a dam to rebuild it. The former dam was built after the manner of dykes, *i.e.*, with upright posts, supporting sleepers, laid inclining up the stream. On these were laid planks, and the planks were covered with dirt. The *Astacus* proceeding up the stream would burrow under the planks where they rested on the bottom of the stream, removing bushels of dirt and gravel in the course of a night. They travel over the dam in their migrations, *often climbing posts* two or three feet high to gain the pond above.*

We have to add a new and eccentric variety of nature—the Pill-making Crab, which abounds at Labuan, Singapore, and Lahore, and is described in Mr. Collingwood's "Rambles of a Naturalist." When the tide is down, this little creature, if stealthily watched, may be seen creeping up a hole in the sandy shore, taking up rapidly particles of the loose powdery sand in its claws, and depositing them in a groove beneath the thorax. A little ball of sand, about the size of a filbert, is forthwith projected, though whether it passes actually through the mouth is not made clear. Pill after pill is seized with one claw, and laid aside, until the beach is covered with these queer little pellets. This is evidently the creature's mode of extracting particles of food from the sand.

Mr. Collingwood also describes, as met with on the shores and waters of the China seas, Glass Crabs, whose flat, transparent, leaf-like bodies seem made

* American Journal of Science and Art.

of fine plates of mica. The dredge brings up many a rich haul of sponges, corals, and gorgoniæ, of the most splendid colours, certain of the sponges harbouring within their cells minute crabs of a new genus. Between Aden and Galle the sea is of a pinkish colour, owing to the immense accumulation of minute kinds of medusæ, in solid masses of red jelly. Over Fiery Cross Reef, the mirror-like sea reveals, at the depth of sixty or seventy feet, this wealth of natural treasures. "Glorious masses of living coral strew the bottom: immense globular madrepores—vast overhanging mushroom-shaped expansions, complicated ramifications of interweaving branches, mingled with smaller and more delicate species—round, finger-shaped, horn-like and umbrella-form—lie in wondrous confusion. Here and there is a large clam-shell, wedged in between masses of coral, the gaping, zigzag mouth covered with the projecting mantle of the deepest Prussian blue; beds of dark purple, long-spined Echini, and the thick black bodies of sea-cucumbers vary the aspect of the sea bottom." *

* W. C. Linnæus Martin, F.L.S.

MUSICAL LIZARDS.

 SMALL Lizard, lately brought home from the Isle of Formosa by Mr. Swinhoe, is decided to be a new species by Dr. Günther, of the British Museum. Mr. Swinhoe found the eggs of this Gecko, or Lizard, in holes of walls or among mortar rubbish. They are round, and usually lie several together, resembling eggs of ordinary Lizards. The young, when first hatched, keep much under stones in dark cellars, where they remain until they attain about two-thirds of the adult size, when they begin to appear in public to catch insects, but evincing great shyness of their seniors. Mr. Swinhoe states that on the plaster-washed sides of his bedroom, close to the angle of the roof, every evening when the lamp was placed on the table below, four little Musical Lizards used to make their appearance and watch patiently for insects attracted by the light. A sphinx or a beetle buzzing into the room would put them into great excitement, and they would run with celerity from one part of the

wall to the other after the deluded insect as it fluttered in vain, buffeting its head, up and down the wall. Two or three would run after the same insect, but as soon as one had succeeded in securing it, the rest would prudently draw aloof. In running over the perpendicular face of the wall they keep so close, and their movements are made so quickly, with one leg in advance of the other, that they have the appearance at a distance of gliding rather than running. The tail is somewhat writhed as the body is jerked along, and much so when the animal is alarmed and doing its utmost to escape; but its progress even then is in short runs, stopping at intervals and raising its head to look about. If a fly perch on the wall it cautiously approaches to within a short distance, then suddenly darts forwards, and with its quickly-protruded, glutinous tongue, fixes it. Apart from watching its curious manœuvres after its insect-food, the attention of the most listless would be attracted by the singular series of loud notes these creatures utter at all hours of the day and night, more especially during cloudy and rainy weather. These notes resemble the syllables "chuck-chuck," several times repeated; and, from their more frequent occurrence during July and August, they are thought to be the call notes of the male to the female.

During the greater part of the day, the little creature lies quiescent in some cranny among the beams of the roof or in the wall of the house, where, however, it is ever watchful for the incautious fly that approaches its den, upon whom it darts forth with but little notice. But it is by no means confined to the habitations of men. Every old wall, and almost every tree, possesses a tenant or two of this species. It is excessively lively, and even when found quietly ensconced in a hole, generally manages to escape—its glittering little eyes (black, with yellow ochre iris) appearing to know no sleep; and an attempt to capture the runaway seldom results in more than the seizure of an animated tail, wrenched off with a jerk by the little fellow as it slips away, without loss of blood. The younger individuals are much darker than the larger and older animals, which are sometimes almost albinoes. In ordinary fly-catching habits, as they stick to the sides of a lamp, there is much similarity between this gecko and the little papehoo, or wall-lizard of China; but this is decidedly a larger and much more active animal, and often engages in a struggle with insects of very large size. The Chinese colonists of Formosa greatly respect the geckos, in consequence of a legend which attributes to them the honour of having once poisoned the supplies of an invading rebellious army, which was thereby totally cut to pieces. The geckos were raised to the rank of generals by the grateful Emperor of China; which honour, the legend states, they greatly appreciated, and henceforth devoted their energies to the extermination of mosquitoes and other injurious insects.

CHAMELEONS, AND THEIR CHANGES.

> "Nil fuit unquam
> Sic impar sibi."—*Horat.*
>
> "Sure such a various creature ne'er was seen."
> *Francis, in imit.*

THE Chameleon tribe is a well-defined family of lizard-like reptiles, whose characters may be summed up as existing in the form of their feet; the toes, which are joined together or bound up together in two packets or bundles, opposed to each other; in their shagreen-like skin; in their prehensile tail; and in their extensile and retractile vermiform tongue.

That the Chameleon was known to the ancients there is no doubt. Its name we derive directly from the *Chamelæo* of the Latins. Aristotle's history of the animal proves the acute observation of that great zoologist—the absence of a sternum, the disposition of the ribs, the mechanism of the tail, the motion of the eyes, the toes bound up in opposable bundles, &c.—though he is not entirely correct on some points. Pliny mentions it, but his account is for the most part a compilation from Aristotle.

Calmet's description of the Chameleon is curiously minute:—"It has four feet, and on each foot three claws. Its tail is long: with this, as well as with his feet, it fastens itself to the branches of trees. Its tail is flat, its nose long, ending in an obtuse point; its back is sharp, its skin plaited, and jagged like a saw, from the neck to the last joint of the tail, and upon its head it has something like a comb; like a fish, it has no neck. Some have asserted that it lives only upon air, but it has been observed to feed on flies, catched with its tongue, which is about ten inches long and three thick, made of white flesh, round, but flat at the end, or hollow and open, resembling an elephant's trunk. It also shrinks, and grows longer. This animal is said to assume the colour of those things to which it is applied; but our modern observers assure us that its natural colour, when at rest, and in the shade, is a bluish-grey; though some are yellow, others green, but both of a smaller kind. When it is exposed to the sun, the grey changes into a darker grey, inclining to a dun colour, and its parts which have least of the light upon them are changed into spots of different colours. Sometimes, when it is handled, it seems speckled with dark spots, inclining to green. If it be put upon a black hat, it appears to be of a violet colour; and sometimes, if it be wrapped up in linen, it is white; but it changes colour only in some parts of the body."

Its changes of colour have been commemorated by the poets. Shakspeare has—

> " I can add colours ev'n to the Chameleon:
> Change shapes with Proteus, for advantage."

Dryden has—

> "The thin Chameleon, fed with air, receives
> The colour of the thing to which it cleaves."

Prior has—

> "As the Chameleon, which is known
> To have no colours of his own,
> But borrows from his neighbour's hue
> His white or black, his green or blue."

Gay, in his charming fable of the Spaniel and the Chameleon, " scarce distinguished from the green," makes the latter thus reply to the taunts of the pampered spaniel :—

> " ' Sir,' says the sycophant, ' like you,
> Of old, politer life I knew:
> Like you, a courtier born and bred,
> Kings lean'd their ear to what I said :
> My whisper always met success ;
> The ladies prais'd me for address ;
> I knew to hit each courtier's passion,
> And flatter'd every vice in fashion :
> But Jove, who hates the liar's ways,
> At once cut short my prosperous days,
> And, sentenced to retain my nature,
> Transform'd me to this crawling creature.
> Doom'd to a life obscure and mean,
> I wander'd in the silvan scene:
> For Jove the heart alone regards ;
> He punishes what man rewards.

How different is thy case and mine!
With men at least you sup and dine;
While I, condemned to thinnest fare,
Like those I flatter'd, fed on air.'"

Upon this fable a commentator acutely notes:—
" The raillery at court sycophants naturally pervades our poet's writings, who had suffered so much from them. Here, however, he intimates something more, namely, the apposite dispensations to man's acts, even in this world. The crafty is taken in by his own guile, the courtier falls by his own arts, and the ladder of ambition only prepares for the aspirant a further fall." *

With respect to the air-food of the Chameleon, Cuvier observes that its lung is so large that, when it is filled with air, it imparts a transparency to the body, which made the ancients say that it lived upon air; and he inclines to think that to its size the Chameleon owes the property of changing its colour; but, with regard to this last speculation, he was wrong, as we shall presently see.

It was long thought that the Chameleon, like most of the lizard tribe, was produced from an egg. The little animal is, however, most clearly viviparous, and not oviparous, although the tales told of the lizard tribe in the story books are most perplexing. To name a few of them:—1. The crocodile, which is the largest of the lizard tribe, and has even attained

* The Fables of John Gay. Illustrated. With Original Memoir, Introduction, and Annotations. By Octavius Freire Owen, M.A., F.S.A. 1854.

the size of 18½ ft. in length, is confidently stated as laying eggs, which she covers with sand and leaves, to be hatched by the sun; and these have been met with in the rivers Nile, Niger, and Ganges. 2. *Lacerta Gangetica,* unknown to Linnæus, but brought to this country from Bengal in 1747 by the late Dr. Mead, is said to be furnished with a false belly, like the opossum, where the young can be received for protection in time of danger. In this case the egg must have been hatched in the belly of the animal, like the viper. 3. The alligator, or American crocodile, lays a vast quantity of eggs in the sand, near the banks of lakes and rivers, and leaves them to be hatched by the sun; and the young are seldom seen. 4. The cayman, or Antilles crocodile, has furnished its eggs to many collections. 5. A salamander was opened by M. Maupertuis, and its belly was found full of eggs; but in " Les Memoires de l'Academie Royale des Sciences " it is stated that, after a similar operation of the kind, " fifty young ones, resembling the parent animal, were found in its womb all alive, and actively running about the room."

The tongue is the chief organ for taking the insects on which the Chameleon lives. By a curious mechanism, of which the tongue-bone is a principal agent, the Chameleon can protrude this cylindrical tongue, which has its tip covered with a glutinous secretion from the sheath at the lower part of the mouth, to the length of six inches. When the Chameleon is about to seize an insect, it rolls round its extraordinary eyeballs so as to bring them to bear on the doomed object; as soon as it arrives within the range of the tongue, that organ is projected with unerring precision, and returns into the mouth with the prey adhering to the viscous tip. The wonderful activity with which this feat is performed, forms a strong contrast to the almost ridiculously slow motions of the animal. Their operation of taking meal-worms, of which they are fond, though comparatively rapid, is not remarkable for its quickness, but done with an act of deliberation, and so that the projection and retraction of the tongue can be very distinctly followed with the eye.

The eyes of the Chameleon are remarkable objects; large, projecting, and almost entirely covered with the shagreen-like skin, with the exception of a small aperture opposite the pupil; their motions are completely independent of each other. It adds to the strange and grotesque appearance of this creature to see it roll one of its eye-globes backwards, while the other is directed forwards, as if making two distinct surveys at one time. Its sight must be acute, from the unerring certainty with which it marks and strikes its prey.

The Chameleons spend their lives in trees, for clinging to the branches of which their organization is admirably adapted. There they lie in wait for the insects which may come within their reach; and it has been thought that, in such situations, their faculty of changing colour becomes highly important in aiding them to conceal themselves.

The powers of abstinence possessed by this singular race are very great; and hence, most probably, arose the old fable of their *living on air*, which was for a long time considered to be "the Chameleon's dish." One has been known to fast upwards of six weeks without taking any sustenance, though meat-food and insects were procured for it. Notwithstanding this fast, it did not appear to fall away much. It would fix itself by the feet and tail to the bars of the fender, and there remain motionless, enjoying the warmth of the fire for hours together. Hasselquist describes one, that he kept for nearly a month, as climbing up and down the bars of its cage in a very lively manner.

The power of the Chameleon's changing colour long exercised the ingenuity of the old naturalists. Hasselquist thought that the changes of colour depended on a kind of disease, more especially a sort of jaundice, to which the animal was subject, particularly when it was put in a rage. M. D'Obsonville thought that he had discovered the secret in the blood, and that the change of colour depended upon a mixture of blue and yellow, whence the different shades of green were derived; and these colours he obtains from the blood and the blood-vessels. Thus he says that the blood is of a violet hue, and will retain its colour on linen or paper for some minutes if previously steeped in a solution of alum, and that the coats of the vessels are yellow; consequently, he argues, that the mixture of the two will produce green. He further traces the change of colour to the passions of the animal. Thus, when a healthy Chameleon is provoked, the circulation is accelerated, the vessels that are spread over the skin are distended, and a superficial blue-green colour is produced. When, on the contrary, the animal is imprisoned, impoverished, and deprived of free air, the circulation becomes languid, the vessels are not filled, the colour of their coats prevails, and the Chameleon changes to a yellow-green, which lasts during its confinement.

Barrow, in his "Travels in Africa," declares that previously to the Chameleon's assuming a change of colour, it makes a long inspiration, the body swelling out to twice its usual size; and as the inflation subsides, the change of colour gradually takes place, the only permanent marks being two small dark lines passing along the sides. Mr. Wood conceives from this account that the animal is principally indebted for these varied tints to the influence of oxygen. Mr. Spittal also regards these changes as connected with the state of the lungs; and Mr. Houston considers this phenomenon as dependent on the turgescency of the skin. Dr. Weissenborn thinks it not unlikely that the nervous currents may directly co-operate in effecting the changes of colour in the Chameleon.

Mr. H. N. Turner, writing from personal observation of the phenomenon in a live Chameleon in his possession, says:—" It has been generally imagined that the purpose of the singular faculty accorded to the Chameleon is to enable it to accommodate its appearance to that of surrounding objects." Mr. Turner's observations do not, however, favour the idea, but seem rather to negative it. The box in which Mr. Turner's Chameleon was kept was of deal, with glass at the top, and a piece of flannel laid at the bottom, a small branching stick being placed there by way of a perch. He introduced, at various times, pieces of coloured paper, covering the bottom of the box, of blue, yellow, and scarlet, but without the slightest effect upon the appearance of the animal. Considering that these primary colours were not such as it would be likely to be placed in contact with in a state of nature, he next tried a piece of green calico, but equally without result. The animal went through all its usual changes without their being in any way modified by the colour placed underneath it. The general tint approximated, as may be readily observed, to those of the branches of trees, just as those of most animals do to the places in which they dwell; but Mr. Turner did not observe the faculty of changing called into play with any apparent object. It is only when the light is removed that the animal assumes a colour which absorbs but little of it.

Not to go further into the numerous treatises which have been published on this intricate subject without arriving at a just conclusion, we refer to the able and interesting paper of Mr. Milne Edwards, for whose acuteness the solution of this puzzling phenomenon was reserved. The steps by which he first overthrew the received theories on the subject, and then arrived at the cause of the change of colour, is shown in the following results, derived from observing two Chameleons living, and researches after the animals had died, on the structure of their skin, and the parts immediately beneath it.

1. That the change in the colour of the Chameleon does not depend essentially either on the more or less considerable swelling of their bodies, or the changes which might hence result to the condition of their blood or circulation; nor does it depend on the greater or less distance which may exist between the several cutaneous tubercles; although it is not to be denied that these circumstances probably exercise some influence upon the phenomenon.

2. That there exist in the skin of these animals two layers of membranous pigment, placed the one above the other, but disposed in such a way as to appear simultaneously under the cuticle, and sometimes in such a manner that the one may hide the other.

3. That everything remarkable in the changes of colour in the Chameleon may be explained by the appearance of the pigment of the deeper layer to an extent more or less considerable, in the midst of the pigment of the superficial layer, or from its disappearance beneath this layer.

4. That these displacements of the deeper pigment do in reality occur; and it is a probable consequence that the Chameleon's colour changes during life, and may continue to change even after death.

5. That there exists a close analogy between the

mechanism by the help of which the change of colour appears to take place in these reptiles, and that which determines the successive appearance and disappearance of coloured spots in the mantles of several of the cephalopods.

Chameleons are found in warm climates of the old world, South of Spain, Africa, East Indies, Isles of Sechelles, Bourbon, France, Moluccas, Madagascar (where it is said there are seven of the species which belong to Africa), Fernando Po, and New South Wales. In the year 1860, a new and curiously formed species of Chameleon was brought from the interior of the Old Calabar district of West Africa, by one of the natives. It is characterized by three horny processes on the head. Many Lizards have singular spiny projections on all parts of the body; but this very well marked species had not been hitherto recorded.

Mrs. Belzoni, the wife of the celebrated traveller in the East, made some careful observations upon the habits of Chameleons, which are worth quoting. The Arabs in Lower Egypt catch Chameleons by jumping upon them, flinging stones at them, or striking them with sticks, which hurts them very much. The Nubians lay them down gently on the ground, and when they come down from the date-trees, they catch hold of the tail of the animal, and fix a string to it; therefore the body does not get injured. Mrs. Belzoni had some Chameleons for several months in her house, and her observations are as follows:—

"In the first place they are very inveterate towards each other, and must not be shut up together, else they will bite each other's tails and legs off.

"There are three species of Chameleons, whose colours are peculiar to themselves: for instance, the commonest sort are those which are generally green, that is to say, the body all green, and, when content, beautifully marked on each side regularly on the green with black and yellow, not in a confused manner, but as if drawn. This kind is in great plenty; they never have any other colour except a light green when they sleep, and when ill, a very pale yellow. Out of near forty I had the first year when in Nubia, I had but one, and that a very small one of the second sort, which had red marks. One Chameleon lived with me eight months, and most of that time I had it fixed to the button of my coat: it used to rest on my shoulder or on my head. I have observed, when I have kept it shut up in a room for some time, that on bringing it out in the air it would begin drawing the air in, and on putting it on some marjorum it has had a wonderful effect on it immediately: its colour became most brilliant. I believe it will puzzle a good many to say what cause it proceeds from. If they did not change when shut up in a house, but only on taking them in a garden, it might be supposed the change of the colours was in consequence of the smell of the plants; but when in a house, if it is watched, it will change every ten minutes: some moments a plain green, at others all its beautiful colours will

CHAMELEONS.

come out, and when in a passion it becomes of a deep black, and will swell itself up like a balloon, and, from being one of the most beautiful animals, it becomes one of the most ugly. It is true that Chameleons are extremely fond of the fresh air, and on taking them to a window when there is nothing to be seen, it is easy to observe the pleasure they certainly take in it: they begin to gulp down the air, and their colour becomes brighter. I think it proceeds, in a great degree, from the temper they are in : a little thing will put them in a bad humour : if in crossing a table, for instance, you stop them, and attempt to turn them another road, they will not stir, and are extremely obstinate: on opening the mouth at them, it will set them in a passion : they begin to arm themselves by swelling and turning black, and will sometimes hiss a little, but not much.

"The third I brought from Jerusalem was the most singular of all the Chameleons I ever had : its temper, if it can be so called, was extremely sagacious and cunning. This one was not of the order of the green kind, but a disagreeable drab, and it never once varied in its colour in two months. On my arrival in Cairo, I used to let it crawl about the room on the furniture. Sometimes it would get down, if it could, and hide itself away from me, but in a place where it could see me; and sometimes, on my leaving the room and on entering it, would draw itself so thin as to make itself nearly on a level with whatever it might be on, so that I might not see it. It had often deceived me so. One day having missed it for some time, I concluded it was hid about the room; after looking for it in vain, I thought it had got out of the room and made its escape: in the course of the evening, after the candle was lighted, I went to a basket that had got a handle across it: I saw my Chameleon, but its colour entirely changed, and different to any I ever had seen before: the whole body, head and tail, a brown with black spots, and beautiful deep orange-coloured spots round the black. I certainly was much gratified. On being disturbed, its colours vanished, unlike the others; but after this I used to observe it the first thing in the morning, when it would have the same colours. Some time after, it made its escape out of my room, and I suppose got into the garden close by. I was much vexed, and would have given twenty dollars to have recovered it again, though it only cost me threepence, knowing I could not get another like it; for, afterwards being in Rosetta, I had between fifty and sixty; but all those were green, yellow, and black; and the Arabs, in catching them, had bruised them so much, that after a month or six weeks they died. It is an animal extremely hard to die. I had prepared two cages with separate divisions, with the intention of bringing them to England; but though I desired the Arabs that used to get them for me to catch them by the tail, they used to hurt them much with their hands; and if once the body is squeezed, it will never live longer than two months. When they used to sleep at night, it was easy to see where they had been bruised; for being of a very light colour when sleeping, the part that had been bruised, either on the body or the head, which was bone, was extremely black, though when green it would not show itself so clear. Their chief food was flies: the fly does not die immediately on being swallowed, for upon taking the Chameleon up in my hands, it was easy to feel the fly buzzing, chiefly on account of the air they draw in their inside: they swell much, and particularly when they want to fling themselves off a great height, by filling themselves up like a balloon: on falling, they get no hurt, except on the mouth, which they bruise a little, as that comes first to the ground. Sometimes they will not drink for three or four days, and when they begin they are about half an hour drinking. I have held a glass in one hand while the Chameleon rested its two forepaws on the edge of it, the two hind ones resting on my other hand. It stood upright while drinking, holding its head up like a fowl. By flinging its tongue out of its mouth the length of its body, and instantaneously catching the fly, it would go back like a spring. They will drink mutton broth: how I came to know this was, one day having a plate of broth and rice on the table where it was: it went to the plate and got half into it, and began drinking, and trying to take up some of the rice, by pushing it with its mouth towards the side of the plate, which kept it from moving, and in a very awkward way taking it into its mouth."

In the autumn of 1868, a pair of Chameleons, in the possession of the Hon. Lady Cust, of Leasowe

Castle, Cheshire, produced nine active young ones, like little alligators, less than an inch long. Such a birth has been, it is believed, very rare in this country. It was remarked, in the above case, that the male and female appeared altogether indifferent about their progeny.

Whatever may be the cause, the fact seems to be certain, that the Chameleon has an antipathy to objects of a black colour. One, which Forbes kept, uniformly avoided a black board which was hung up in the chamber; and, what is most remarkable, when the Chameleon was held forcibly before the black board, it trembled violently and assumed a *black colour.**

It may be something of the same kind which makes Bulls and Turkey-cocks dislike the colour of scarlet, a fact of which there can be no doubt.

* This, it will be seen by referring to page 307, does not correspond with Calmet's statement.

RUNNING TOADS.

THAT the Toad, by common repute "ugly and venomous," should be made a parlour pet, is passing strange; yet such is the case, and we find in a letter from Dr. Husenbeth, of Cossey, the following curious instances. Thus he describes a species, there often met with, the eyes of which have the pupil surrounded with bright golden-yellow, whereas in the common toad the circle is red or orange. This remarkable peculiarity Dr. H. has not seen anywhere noticed. The head is like that of the common sort, but much more blunt, and rounded off at the nose and mouth, and the arches over the eyes are more prominent. The most remarkable difference is a line of yellow running all down the back. Also down each side this Toad has a row of red pimples, like small beads, which are tolerably regular, but appear more in some specimens than in others. The general colour is a yellowish-olive, but the animal is beautifully marked with black spots, very regularly disposed, and exactly corresponding on each side of the yellow line down the back. Like all other Toads, this one occasionally changes its colour, becoming more brown, or ash-colour, or reddish at times, probably in certain states of the weather. This species is much more active than the common Toad. It never leaps, and very seldom crawls, but makes a short run, stops a little, and then runs on again. If frightened or pursued, it will run along much quicker than one would suppose.

During the previous summer Dr. H. kept three Toads of this kind in succession. "The first (says Dr. H.) I procured in July; but after a few days, when I let him have a run on the carpet of my parlour, he got into a hole in a corner of the floor, of which I was not aware, and fell, as I suppose, underneath the floor, into the hollow space below. I concluded that he could never get up again, and gave him up to his fate. I then began to keep another Running Toad, which fed well at first, but after three weeks refused food, and evidently wasted; so I turned him out into the garden, and have not met with him since. After more than three weeks, the former Toad reappeared, but how he came up from beneath the floor I never could conceive, or how he had picked up a living in the meantime. He was, however, in good condition, and seemed to have lived well, probably on spiders and woodlice. He had been seen by a servant running about the carpet, but I knew nothing of his having come forth again, till in the evening, when he had got near the door, and it was suddenly opened so as to pass over the poor creature, and crush it terribly. I took it up apparently dead. It showed no sign of life; the eyes were closed, it did not breathe, and the backbone seemed quite broken, and the animal was crushed almost flat. I found a very curious milky secretion exuding from it, where it had been most injured and the skin was most broken. This was perfectly white, and had exactly the appearance of milk thrown over the toad. It did not bleed, though much lacerated; but instead of blood appeared this milky fluid, which had an odour of a most singular kind, different from anything I ever smelt. It is impossible to describe it. It was not fetid, but of a sickly, disgusting, and overpowering character, so that I could not endure to inhale it for a moment. I had read and seen a good deal of the extraordinary powers of revivification in toads, but was not prepared for what I witnessed on this occasion. I laid this poor animal, crushed, flattened, motionless, and to all appearance dead, upon a cold iron plate of the fireplace. He fell over on one side, and showed no sign of life for a full hour. After that he had slightly moved one leg, and so remained for about another half-hour. Then he began to breathe feebly, and gathered up his legs, and his back began to rise up into its usual form. In about two hours from the time of the accident, he had so far recovered as to crawl about, though with difficulty. The milky liquor was reabsorbed, and gradually disappeared as the toad recovered. The next morning it was all gone, and no mark of injury could be seen, except a small hole in his back, which soon closed. He recovered so far as to move about pretty well, but his back appeared to have been broken, and one foreleg crippled. I therefore thought it best to give him his liberty in the garden. But so wonderful and speedy a recovery I could never have believed without ocular testimony.

"I then tried my third and last Running Toad. I began to keep him on Sept. 13th. He was a very fine specimen, and larger than the two former. He fed well, and amused me exceedingly. He was very tame, and would sit on my hand quite quiet, and enjoy my stroking him gently down his head and back. Soon after I got him he began to cast his skin. I helped him to get rid of it by stripping it down each side, which he seemed to like much, and sat very quiet during the operation. The new skin was quite beautiful, and shone as if varnished. This Toad lived in a crystal palace, or glass jar, where I had kept all the others before him. He took food freely, and his appetite was so good that in one day he eat seven large flies and three bees without stings. He was particularly fond of woodlice and earwigs, but would take centipedes, moths, and even butterflies. Being more active than common Toads, he often made great efforts to get out of his glass jar. I used to let him run about the room nearly every day for a short time, and often treated him to a run in the garden. Toads make a slight noise sometimes in the evenings, uttering a short sound like 'coo,' but I never heard them croak. Before wet weather, and during its continuance, my Toad was disinclined for food, and took no notice of flies even walking over his nose. He would then burrow and hide himself in the moss at the bottom of his glass palace. Thus I kept him, and found him very tame and amusing. But after about two months he became more impatient of confinement, and refused to take any food. I did not perceive that he fell away, though his feet and toes turned of a dark colour, which I knew was a sign of being out of condition; and, on the 10th of November, I found him dead. I have now tried three of this sort, and have come to the conclusion that the Running Toad will not live in captivity. This I much regret, as its habits are interesting, and its ways very amusing.

"F. C. HUSENBETH, D.D."

FROG AND TOAD CONCERTS.

It would be hard to believe the stories of the vocal powers of Frogs and Toads were they not related by trustworthy travellers, who tell of animal concerts,

"Wild as the marsh, and tuneful as the harp."

Mr. Priest, the traveller in America, who was himself a musician, records :—" Prepared as I was to hear something extraordinary from these animals,

I confess the first *Frog Concert* I heard in America was so much beyond anything I could conceive of the power of these musicians, that I was truly astonished. This performance was *al fresco*, and took place on the eighteenth of April, in a large swamp, where there were at least 10,000 performers; and I really believe not two exactly in the same pitch, if the octave can possibly admit of so many divisions, or shades of semitones."

Professor and Mrs. Louis Agassiz, in their recent " Journey in Brazil," record:—" We must not leave Parà without alluding to our evening concerts from the adjoining woods and swamps. When I first heard this strange confusion of sounds, I thought it came from a crowd of men shouting loudly, though at a little distance. To my surprise, I found that the rioters were the frogs and toads in the neighbourhood. I hardly know how to describe this Babel of woodland noises; and, if I could do it justice, I am afraid my account would hardly be believed. At moments it seems like the barking of dogs, then like the calling of many voices on different keys; but all loud, rapid, excited, full of emphasis and variety. I think these frogs, like ours, must be silent at certain seasons of the year, for on our first visit to Parà we were not struck by this singular music, with which the woods now resound at nightfall."

SONG OF THE CICADA.

THE Greeks have been scoffed at for rendering in deathless verse the song of so insignificant an insect as the Cicada; and hence it has been asserted that their love for such slender music must have been either exaggerated or simulated. It is pleasant, however, to hear an independent observer in the other hemisphere confirm their testimony. Mr. Lord tells us that in British Colombia there is one sound or song which is clearer, shriller, and *more singularly tuneful than any other.* It never appears to cease, and it comes from everywhere—from the tops of the trees, from the trembling leaves of the cotton-wood, from the stunted under-brush, from the flowers, the grass, the rocks and boulders—nay, the very stream itself seems vocal with hidden minstrels, all chanting the same refrain.

An especial feature of the Cicada's song is, that it increases in intensity when the sun is hottest; and one of the later Latin poets mentions the time when its music is at its highest, as an alternative expression for noon. Mr. Tennyson, inadvertently, speaks in "Ænone" of the Grasshopper being silent in the grass, and of the Cicada sleeping when the noonday quiet holds the hill. Keats sings more truly:—

"When all the birds are faint with the hot sun,
And hide in cooling trees, a voice will run
From hedge to hedge about the new-mown mead:
That is the Grasshopper's."

Then the Greek poets show us how intimately the song of the Cicada is associated with the hottest hours of the day. Aristophanes describes it as mad for the love of the sun; and Theocritus, as scorched by the sun. When all things are parched with the heat (says Alcæus), then from among the leaves issues the song of the sweet Cicada. His shrill melody is heard in the full glow of noontide, and the vertical rays of a torrid sun fire him to sing. Over and over again Mr. Lord met with allusions to the same peculiarity.

Cicadæ are regularly sold for food in the markets of South America. They are not eaten now, like they were at Athens, as a whet to the appetite; but they are dried in the sun, powdered, and made into a cake.

STORIES ABOUT THE BARNACLE GOOSE.

"As barnacles turn Poland geese
In th' islands of the Orcades."—*Hudibras.*

ONE of the earliest references to this popular error is in the "Natural Magic" of Baptista Porta, who says:—"Late writers report that not only in Scotland, but also in the river of Thames by London, there is a kind of shell-fish in a two-leaved shell, that hath a foot full of plaits and wrinkles. . . . They commonly stick to the keel of some old ship. Some say they come of worms, some of the boughs of trees which fall into the sea; if any of them be cast upon shore, they die; but they which are swallowed still into the sea, live and get out of their shells, and grow to be ducks, or such-like birds."

Professor Max Müller, in a learned lecture, enters fully into the origin of the different stories about the Barnacle Goose. He quotes from the "Philosophical Transactions" of 1678 a full account by Sir Robert Moray, who declared that he had seen within the barnacle shell, as through a concave or diminishing glass, the bill, eyes, head, neck, breast, wings, tail, feet, and feathers of the Barnacle Goose. The next witness was John Gerarde, Master in Chirurgerie, who, in 1597, declared that he had seen the actual metamorphosis of the muscle into the bird, describing how—

"The shell gapeth open, and the first thing that appeareth is the fore said lace or string; next come the leg of the birde hanging out, and as it groweth greater, it openeth the shell by degrees, till at length it is all come forth, and hangeth only by the bill, and falleth into the sea, when it gathereth feathers and groweth to a foule, bigger than a mallart; for the truth hereof, if any doubt, may it please them to repair unto me, and I shall satisfie them by the testimonies of good witnesses."

As far back as the thirteenth century, the same story is traced in the writings of Giraldus Cambrensis. This great divine does not deny the truth of the miraculous origin of the Barnacle Geese, but he warns the Irish priests against dining off them during Lent on the plea that they were not flesh, but fish. For, he writes, "If a man during Lent were to dine off a leg of Adam, who was not born of flesh either, we should not consider him innocent of having eaten what is flesh." This modern myth, which, in spite of the protests of such men as Albertus Magnus, Æneas Sylvius, and others, maintained its ground for many centuries, and was defended, as late as 1629, in a book by Count Maier, "De volucri arborea," with arguments, physical, metaphysical, and theological, owed its origin to a play of words. The muscle shells are called *Bernaculæ* from the Latin *perna*, the mediæval Latin *berna*; the birds are called *Hibernicæ* or *Hiberniculæ*, abbreviated to *Berniculæ*. As their names seem one, the creatures are supposed to be one, and everything conspires to confirm the first mistake, and to invest what was originally a good Irish story—a mere *canard*—with all the dignity of scientific, and all the solemnity of theological truth. The myth continued to live until the age of Newton. Specimens of *Lepadidæ*, prepared by Professor Rolleston of Oxford, show how the outward appearance of the *Anatifera* could have supported the popular superstition which derived the *Bernicla*, the goose, from the *Bernicula*, the shell.

Drayton (1613), in his "Poly-olbion," iii., in connexion with the river Lee, speaks of

"Th' anatomised fish and fowls from planchers sprung;"

to which a note is appended in Southey's edition, p. 609, that such fowls were "Barnacles, a bird breeding upon old ships." A bunch of the shells attached to the ship, or to a piece of floating timber, at a distance appears like flowers in bloom; the foot of the animal has a similitude to the stalk of a plant growing from the ship's sides, the shell resembles a calyx, and the flower consists of the tentacula, or fingers, of the shell-fish. The ancient error was to mistake the foot for the neck of a goose, the shell for its head, and the tentacula for feathers. As to the body, *non est inventus*.

Sir Kenelm Digby was soundly laughed at for relating to a party at the castle of the Governor of Calais, that "the Barnacle, a bird in Jersey, was first a shell-fish to appearance, and, from that striking upon old wood, became in time a bird." In 1807, there was exhibited in Spring-gardens, London, a "Wonderful natural curiosity, called the Goose Tree, Barnacle Tree, or Tree bearing Geese," taken up at sea on January 12th, and more than twenty men could raise out of the water.*

Sir J. Emerson Tennent asks whether the ready acceptance and general credence given to so obvious a fable may not have been derived from giving too literal a construction to the text of the passage in the first chapter of Genesis:—

"And God said, Let the *waters bring forth abundantly* the moving creature that hath life, and the *fowl* that may fly in the open firmament of heaven."

The Barnacle Goose is a well-known bird, and is eaten on fast-days in France, by virtue of this old belief in its marine origin. The belief in the barnacle origin of the bird still prevails on the west coast of Ireland, and in the Western Highlands of Scotland.

The finding of the Barnacle is thus described by Mr. Sidebotham, to the Microscopical and Natural History Section of the Literary and Philosophical Society:—"In September, I was at Lytham with my family. The day was very stormy, and the previous night there had been a strong south-west wind,

* "Notes and Queries," No. 201.

and evidences of a very stormy sea outside the banks. Two of my children came running to tell me of a very strange creature that had been washed up on the shore. They had seen it from the pier, and pointed it out to a sailor, thinking it was a large dog with long hair. On reaching the shore I found a fine mass of Barnacles, *Pentalasinus anatifera*, attached to some staves of a cask, the whole being between four and five feet long. Several sailors had secured the prize, and were getting it on a truck to carry it away. The appearance was most remarkable, the hundreds of long tubes with their curious shells looking like what one would fancy the fabled Gorgon's head with its snaky locks. The curiosity was carried to a yard where it was to be exhibited, and the bellman went round to announce it under the name of the sea-lioness, or the great sea-serpent. Another mass of Barnacles was washed up at Lytham, and also one at Blackpool, the same day or the day following. This mass of Barnacles was evidently just such a one as that seen by Gerard at the Pile of Foulders. It is rare to have such a specimen on our coasts. The sailors at Lytham had never seen anything like it, although some of them were old men who had spent all their lives on the coast."

LEAVES ABOUT BOOKWORMS.

ON paper, leather, and parchment are found various animals, popularly known as "Bookworms." Johnson describes it as a worm or mite that eats holes in books, chiefly when damp; and in the "Guardian" we find this reference to its habits:—"My lion, like a moth or bookworm, feeds upon nothing but paper."

Many years ago an experienced keeper of the Ashmolean Museum at Oxford collected these interesting details of Bookworms:—"The larvæ of *Crambus pinguinalis* will establish themselves upon the binding of a book, and spinning a robe will do it little injury. A mite, *Acarus eruditus*, eats the paste that fastens the paper over the edges of the binding and so loosens it. The caterpillar of another little moth takes its station in damp old books, between the leaves, and there commits great ravages. The little boring wood-beetle, who attacks books and will even bore through several volumes. An instance is mentioned of twenty-seven folio volumes being perforated in a straight line, by the same insect, in such a manner that by passing a cord through the perfect round hole made by it the twenty-seven volumes could be raised at once. The wood-beetle also destroys prints and drawings, whether framed or kept in a portfolio."

There is another "Bookworm," which is often confounded with the Death-watch of the vulgar; but is smaller, and instead of beating at intervals, as does the Death-watch, continues its noise for a considerable length of time without intermission. It is usually found in old wood, decayed furniture, museums, and neglected books. The female lays her eggs, which are exceedingly small, in dry, dusty places, where they are least likely to meet with disturbance. They are generally hatched about the beginning of March, a little sooner or later, according to the weather. After leaving the eggs, the insects are so small as to be scarcely discerned without the use of a glass. They remain in this state about two months, somewhat resembling in appearance the mites in cheese, after which they undergo their change into the perfect insect. They feed on dead flies and other insects; and often, from their numbers and voracity, very much deface cabinets of natural history. They subsist on various other substances, and may often be observed carefully hunting for nutritious particles amongst the dust in which they are found, turning it over with their heads, and searching about in the manner of swine. Many live through the winter buried deep in the dust to avoid the frost.

The best mode of destroying the insects which infest books and MSS. has often occupied the attention of the possessors of valuable libraries. Sir Thomas Phillips found the wood of his bookcase attacked, particularly where beech had been introduced, and appeared to think that the insect was much attracted by the paste employed in binding. He recommended as preservatives against their attacks spirits of turpentine and a solution of corrosive sublimate, and also the latter substance mixed with paste. In some instances he found the produce of a single impregnated female sufficient to destroy a book. Turpentine and spirit of tar are also recommended for their destruction; but the method pursued in the collections of the British Museum is an abundant supply of camphor, with attention to keeping the rooms dry, warm, and ventilated. Mr. Macleay states it is the *acari* only which feed on the paste employed in binding books, and the larvæ of the Coleoptera only which pierce the boards and leaves.

The ravages of the Bookworm would be much more destructive had there not been a sort of guardian to the literary treasures in the shape of a spider, who, when examined through a microscope, resembles a knight in armour. This champion of the library follows the Worm into the book-case, discovers the pit he has digged, rushes on his victim, which is about his own size, and devours him. His repast finished, he rests for about a fortnight, and when his digestion is completed, he sets out to break another lance with the enemy.

The Death-watch, already referred to, and which must be acquitted of destroying books, is chiefly known by the noise which he makes behind the wainscoting, where he ticks like a clock or watch. How so loud a noise is produced by so small an insect has never been properly explained; and the ticking has led to many legends. The naturalist Degeer relates that one night, in the autumn of 1809, during an entomological excursion in Brittany, where travellers were scarce and accommodation bad, he sought hospitality at the house of a friend. He was from home, and Degeer found a great deal of trouble in gaining admittance; but at last the peasant who had charge of the house told Degeer that he would give him "the chamber of death," if he liked. As Degeer was much fatigued, he accepted the offer. "The bed is there," said the man, "but no one has slept in it for some time. Every night the spirit of the officer, who was surprised and killed in this room by some chouans, comes back. When the officer was dead, the peasants divided what he had about him, and the officer's watch fell to my uncle, who was delighted with the prize, and brought it home to examine it. However, he soon found out that the watch was broken, and would not go. He then placed it under his pillow, and went to sleep; he awoke in the night, and to his terror heard the ticking of a watch. In vain he sold the watch, and gave the money for masses to be said for the officer's soul, the ticking continued, and has never ceased." Degeer said that he would exorcise the chamber, and the peasant left him, after making the sign of the cross. The naturalist at once guessed the riddle, and, accustomed to the pursuit of insects, soon had a couple of Death-watches shut up in a tin case, and the ticking was reproduced.

Swift has prescribed this destructive remedy by way of ridicule :—

"A Wood-worm
That lies in old wood, like a hare in her form :
With teeth or with claws it will bite, or will scratch ;
And chambermaids christen this worm a Death-watch,
Because like a watch it always cries click :
Then woe be to those in the house that are sick !
For, sure as a gun, they will give up the ghost
If the maggot cries click when it scratches the post.
But a kettle of scalding hot water ejected,
Infallibly cures the timber affected :
The omen is broken, the danger is over ;
The maggot will die, the sick will recover."

BORING MARINE ANIMALS, AND HUMAN ENGINEERS.

WERE a young naturalist asked to exemplify what man has learned from the lower animals, he could scarcely adduce a more striking instance than that of a submarine shelly worker teaching him how to execute some of his noblest works. This we have learned from the life and labours of the *Pholas*, of which it has been emphatically said:—" Numerous accounts have been published during the last fourteen years in every civilized country and language of the boring process of the *Pholas*; and machines formed on the model of its mechanism have for years been tunnelling Mont Cenis."

In the Eastern Zoological Gallery of the British Museum, cases 35 and 36, as well as in the Museum of Economic Geology in Piccadilly, may be seen specimens of the above very curious order of *Conchifers*, most of the members of which are distinguished by their habits of boring or digging, a process in which they are assisted by the peculiar formation of the foot, from which they derive their name. Of these ten families one of the most characteristic is that of the Razor-shells, which, when the valves are shut, are of a long, flattened, cylindrical shape, and open at both ends. Projecting its strong pointed foot at one of these ends, the *solen* can work itself down into the sand with great rapidity, while at the upper end its respiratory tubes are shot out to bring the water to its gills. Of the *Pholadæ*, the shells of which are sometimes called multivalve, because, in addition to the two chief portions, they have a number of smaller accessory pieces, some bore in hard mud, others in wood, and others in rocks. They fix themselves firmly by the powerful foot, and then make the shell revolve; the sharp edges of this commence the perforation, which is afterwards enlarged by the rasp-like action of the rough exterior; and though the shell must be constantly worn down, yet it is replaced by a new formation from the animal, so as never to be unfit for its purpose. The typical bivalve of this family is the *Pholas*, which bores into limestone-rock and other hard material, and commits ravages on the piers, breakwaters, &c., that it selects for a home.

In the same family as the above Dr. Gray ranks the *Teredo*,* or wood-boring mollusc, whose ravages on ships, piles, wooden piers, &c., at sea resemble those of the white ant on furniture, joints of houses, &c., on shore. Perforating the timber by exactly the same process as that by which the Pholas per-

* How Brunel took his construction of the Thames Tunnel from observing the bore of the *Teredo navalis* in the keel of a ship, in 1814, is well known.

forates the stones, the Teredo advances continually, eating out a contorted tube or gallery, which it lines behind it with calcareous matter, and through which it continues to breathe the water.

The priority of the demonstration of the Pholas and its "boring habits" has been much disputed. The evidence is full of curious details. It appears that Mr. Harper, of Edinburgh, author of "The Sea-side and Aquarium," having claimed the lead, Mr. Robertson, of Brighton, writes to dispute the originality; adding that he publicly exhibited Pholades in the Pavilion at Brighton in July, 1851, perforating chalk rocks by the raspings of their valves and squirtings of their syphons. Professor Flourens (says Mr. Robertson) taught my observations to his class in Paris in 1853; I published them in 1851, and again more fully in the "Journal de Conchyliologie," in 1853; and M. Emile Blanchard illustrated them in the same year in his "Organisation du Règne Animal." I published a popular account of the perforating processes in "Household Words" in 1856. After obtaining the suffrages of the French authorities, I have been recently honoured with those of the British naturalist. (See Woodward's "Recent and Fossil Shells," p. 327. Family, Pholadidæ.) On returning to England last autumn I exhibited perforating Pholades to all the naturalists who cared to watch them. An intelligent lady whom I supplied with Pholades has made a really new and original observation, which I may take this opportunity of communicating to the public. She observed two

Pholades whose perforations were bringing them nearer and nearer to each other. Their mutual raspings were wearing away the thin partition which separated their crypts. She was curious to know what they would do when they met, and watched them closely. When the two perforating shellfish met and found themselves in each other's way, the stronger just bored right through the weaker Pholas.*

Mr. Robertson has communicated to "Jameson's Journal," No. 101, the results of his opportunities of studying the Pholas, during six months, to discover how this mollusc makes its hole or crypt in the chalk: by a chemical solvent? by absorption? by ciliary currents? or by rotatory motions? Between twenty and thirty of these creatures were at work in lumps of chalk, in sea-water, in a finger-glass, and open for three months; and by watching their operations, Mr. Robertson became convinced that the Pholas makes its hole by grating the chalk with its rasp-like valves, licking it up when pulverized with its foot, forcing it up through its principal orbrambial syphon, and squirting it out in oblong nodules. The crypt protects the Pholas from *confervæ*, which, when they get at it, grow not merely outside, but even with the lips of the valves, preventing the action of the syphons. In the foot there is a gelatinous spring or style, which, even when taken out, has great elasticity, and which seems the mainspring of the motions of the Pholas.

Upon this Dr. James Stark, of Edinburgh, writes:

* "Athenæum," No. 1640.

—"Mr. Robertson, of Brighton, claims the merit of teaching that Pholades perforate rocks by 'the rasping of their valves and the squirting of their syphons.' His observations only appear to reach back to 1851. But the late Mr. John Stark, of Edinburgh, author of the 'Elements of Natural History,' read a paper before the Royal Society of Edinburgh, in 1826, which was printed in the Society's 'Transactions' of that year, in which he demonstrated that the Pholades perforate the shale rocks in which they occur on this coast, by means of the rasping of their valves, and not by acids or other secretions. From also finding that their shells scratched limestone without injury to the fine rasping rugosities, he inferred that it was by the same agency they perforated the hard limestone rocks."

To this Mr. Robertson replies, that Mr. Osler also, in 1826, demonstrated that the Pholades "perforate the shale rocks by means of the rasping of their valves; and more, for he actually witnessed a rotatory movement. But Réaumur and Poli had done as much as this in the eighteenth and Sibbald in the seventeenth century: and yet I found the solvent hypothesis in the ascendant among naturalists in 1835, when I first interested myself in the controversy. What I did in 1851 was, I exhibited Pholades at work perforating rocks, and explained how they did it. What I have done is, I have made future controversy impossible, by exhibiting the animals at work, and by discovering the anatomy and the physiology of the perforating instruments.

In the words of M. Flourens, 'I made the animals work before my eyes,' and I 'made known their mechanism.' The discovery of the function of the hyaline stylet is not merely a new discovery, it is the discovery of a kind of instrument as yet unique in physiology."

Mr. Harper having termed the boring organ of the Pholas the "hyaline stylet," found it to have puzzled some of the disputants, whereupon Mr. Harper writes:—"Its use up to the present time has been a mystery, but the general opinion of authors seems to be, that it is the gizzard of the Pholas. This I very much doubt, for it is my belief that the presence of such an important muscle is solely for the purpose of aiding the animal's boring operations. Being situated in the centre of the foot, we can readily conceive the great increase of strength thus conveyed to the latter member, which is made to act as a powerful fulcrum, by the exercise of which the animal rotates —and at the same time presses its shell against and rasps the surface of the rock. The question being asked, 'How can the stylet be procured to satisfy curiosity?' I answer, by adopting the following extremely simple plan. Having disentombed a specimen, with the point of a sharp instrument cut a slit in the base of its foot, and the object of your search will be distinctly visible in the shape of, if I may so term it, an opal cylinder. Sometimes I have seen the point of this organ spring out beyond the incision, made as above described."

Lastly, Mr. Harper presented the Editor of the

"Athenæum" with a piece of bored rock, of which he has several specimens. He adds, "On examination, you will perceive that the larger Pholas must have bored through its smaller and weaker neighbour (how suggestive!), the shell of the latter, most fortunately, remaining in its own cavity."

Now, Mr. Robertson claimed for his observation of this phenomenon novelty and originality; but Mr. Harper stoutly maintained it to be " as common to the eye of the practised geologist as rain or sunshine." The details are curious; though some impatient, and not very grateful reader, may imagine himself in the condition of the shell of the smaller Pholas, and will be, as he deserves to remain, in the minority.*

It may be interesting to sum up a few of the opinions of the mode by which these boring operations are performed. Professor Forbes states the mode by which Molluscs bore into wood and other materials is as follows:—" Some of the Gauterspods have tongues covered with silica to enable them to bore, and it was probably by some process of this kind that all the Molluscs bored."

Mr. Peach never observed the species of Pholas to turn round in their holes, as stated by some observers, although he had watched them with great attention. Mr. Charlesworth refers to the fact that, in one species of shell, not only does the hole in the

* See also "Life in the Sea," in "Strange Stories of the Animal World," by the author of the present volume. Second Edition. 1868.

rock which the animal occupies increase in size, but also the hole through which it projects its syphons.

Professor John Phillips, alluding to the theories which have been given of the mode in which Molluscs bore into the rocks in which they live, believes that an exclusively mechanical theory will not account for the phenomenon; and he is inclined to adopt the view of Dr. T. Williams—that the boring of the Pholades can only be explained on the principle which involves a chemical as well as a mechanical agency.

Mr. E. Ray Lankester notices that the boring of Annelids seems quite unknown; and he mentions two cases, one by a worm called Leucadore, the other by a Sabella. Leucadore is very abundant on some shores, where boulders and pebbles may be found worm-eaten and riddled by them. Only stones composed of carbonate of lime are bored by them. On coasts where such stones are rare, they are selected, and others are left. The worms are *quite soft*, and armed only with horny bristles. *How, then, do they bore?* Mr. Lankester maintains that it is by carbonic acid and other acid excretions of their bodies, *aided* by the mechanical action of their bristles. The selection of a material soluble in these acids is most noticeable, since the softest chalk and the hardest limestone are bored with the same facility. This can only be by chemical action. If, then, we have a case of chemical boring in these worms, is it not probable that many Molluscs are similarly assisted in their excavations?

INDEX.

ANCIENT Zoological Gardens, 12
Animals, Rare, of London Zoological Society, 16, 17, 18
Annelids, boring, 348
Annelids and Molluscs, Boring Habits of, 348
Ant-Bear in captivity, 76
Ant-Bear, the Great, 72
Ant-Bear at Madrid, 72
Ant-Bear described, 77
Ant-Bear, Domestic, in Paraguay, 75
Ant-Bear, Economy of, 76
Ant-Bear and its Food, 74
Ant-Bears, Fossil, 80, 81
Ant-Bear, Muscular Force of, 79
Ant-Bear, Wallace's Account of, 73
Ant-Bear, Zoological Society's, 76, 82, 84
Ant-Eater, Porcupine, 84
Ant-Bear, Professor Owen on, 80
Ant-Eaters, scarcity of, 80
Ant-Eater, Tamandua, 82
Ant-Eaters, Von Saek's Account of, 83
Aristotle's History of Animals, 279, 280

BARNACLE GEESE, finding of the, 334
Barnacle Goose, Gerarde on, 332
Barnacle Goose, Giraldus Cambrensis on, 332
Barnacle Goose, Max Müller on, 331
Barnacle Goose, name of, 332
Barnacle Goose, Sir E. Tennent on, 334
Barnacle Goose, Sir Kenelm Digby on, 334
Barnacle Goose, Sir R. Moray on, 331
Barnacle Goose, Stories of the, 331-335
Barnacles breeding upon old ships, 333
Barnacle Geese in the Thames, 331
Bat, altivolans, by Gilbert White, 100
Bat, American, by Lesson, 91
Bat, Aristotle on, 85
Bat, Mr. Bell on, 86
Bats, Curiosities of, 85
Bat, described by Calmet, 87
Bat, Flight and Wing of, 96
Bats, in England, 100

Bat, Heber, Stedman, and Waterton on, 91
Bats in Jamaica, 100
Bat, Kalong, of Java, 93
Bat, Long-Eared, by Sowerby, 92, 93-96
Bat, Nycteris, 97
Bat, Reremouse and Flittermouse, 86
Bat Skeleton, Sir C. Bell on, 87
Bat in Scripture, 85
Bat, Vampire, from Sumatra, 88
Bat, Vampire, Lines on, by Byron, 89
Bat, vulgar errors respecting, 97
Bat-Fowling or Bat-Folding, 92
Berlin Zoological Gardens and Museum, 16
Bible Natural History, 11
Birds, Addison on their Nests and Music, 156, 157
Bird, Australian Bower, Nest of, 167
Bird, Baya, Indian, Nest of, 164
Birds and Animals, Beauty in, 150
Birds, Brain of, 154
Birds, Characteristics of, 145
Birds, Colour of, 148
Bird Confinement, Dr. Livingstone on, 169
Birds' Eggs, large, 162
Birds' Eggs, Colours of, 158
Birds' Eggs and Nests, 158
Birds, European, list of, 161
Birds, Flight of, 146, 147
Birds, Insectivorous, 151; Instinct, Intelligence, and Reason, 217
Bird-Life, 145
Bird-Murder, wanton, 152
Birds' Nesting, 159
Birds' Nests—Cape Swallows, 168
Birds' Nests—Brush Turkey, 171
Birds' Nests, large, 164
Birds' Eggs—Ostrich and Epyornis, 162, 163
Birds' Nests—Tailor Birds, 165-167
Birds, Rapid Flight of, 147
Birds, Signal of Danger among, 155
Birds, Song of, 149
Birds, Mr. Wolley's Collections, 159, 160

Bookworms, Leaves about, 336
Bookworms and Death-watch, 337
Boring Marine Animals, and Human Engineers, 341

CHAMELEON of the Ancients, 306
Chameleon's antipathy to black, 322
Chameleons, Mrs. Belzoni's, 316-320
Chameleons, Birth of, in England, 321
Chameleon changing Colour, 311, 316
Chameleon, Cuvier on, 309
Chameleon, described by Calmet, 307
Chameleon Family, 307
Chameleon, Air-food of, 309
Chameleon, Milne Edwards on its Change of Colour, 314-316
Chameleons, Native Countries of, 316
Chameleon of the Poets, 308
Chameleons, Reproduction of, 309
Chameleon, Tongue and Eyes of, 310, 311
Chinese Zoological Gardens, 12
Cicada, Song of the, 329
Cormorant's Bone, curious, 204
Cormorants, Chase of, 203
Cormorant Fishery in China, 202
Cormorant, Habits of the, 201
Cormorant trained for Fishing, 201
Curiosities of Zoology, 11

ECCENTRICITIES of Penguins, 188; Darwin, Mr., his account of Falkland Islands Penguin, 192; Dassent Island Penguins, 188; Death-watch and Bookworm, 337, 338; Falkland Islands Penguins, 189; King Penguins, 191; Patagonian Penguins, 189; Penguin, the name, 194; Webster, Mr., his Account of Penguins, 193
Epicure's Ortolan, the, 172
Epicurism Extravagant, 177
Evelyn and St. James's Physique Garden, 15

FISH in British Colombia, 280; Candle-fish, 282; Octopus, 283; Salmon Army, 281; Spoonbill Sturgeon, 285; Sturgeons, and Sturgeon Fishing, 284—287
Fish-Talk, 250; Affection of Fishes, 256; Bohemian Wels Fish, 270; Bonita and Flying Fish, 263; Californian Fish, 268; Carp at Fontainebleau, 254; Cat-fish, curious Account of, 257; Double Fish, 272; Fish changing Colour, 251; Fish Noise, 252; Gold Fish, 274; Grampus, gambols of, 262; Great General of the South Sea, 272; Grouper, the, 272; Hassar, the, 256 Hearing of Fishes, 253; Herring Puzzle, 278; Jaculator Fish of Java, 264; Jamaica, Curious Fish at, 266; Little Fishes the Food of Larger, 259; Marine Observatory, 276; Mecho of the Danube, 270; Migration of Fishes, 260; Miller's Thumb, 276; Numbers, vast, of Fishes, 258; Pike, Wonderful, 269; Pilot Fish, 267; Sharks, 267; Singing Fish, 252; Square-browed Malthe, 274; Strange Fishes, 251; Sun-fish, 271; Swimming of Fishes, 250; Swordfish, 266; Warrior Fish, 266
Frog and Toad Concerts, 327

HEDGEHOG, the, 102
Hedgehog devouring Snakes, 104
Hedgehog, Food of, 103
Hedgehogs. Gilbert White on, 107
Hedgehog and Poisons, 105
Hedgehogs, Sir T. Browne on, 102
Hedgehog Sucking Cows, 104
Hedgehog and Viper, Fight between, 106, 107
Hedgehog, Voracity of, 103
Hippopotamus, Ancient History of, 119
Hippopotamus, described by Aristotle and Herodotus, 121
Hippopotamus, Economy of the, 115
Hippopotamus, the, in England, 108
Hippopotami, Fossil, 122
Hippopotami on the Niger, 117
Hippopotamus, Professor Owen's Description of, 111—115
Hippopotamus and River Horse, 116
Hippopotamus in Scripture, 120
Hippopotamus, Utility of, 118
Hippopotamus from the White Nile, 109
Hippopotamus, Zoological Society's, in 1850, 108—111

LEAVES about Bookworms, 336
Lions in Algeria, and Jules Gerard, 143
Lion, African, 131
Lion, Bengal, 133
Lion described by Bennett, 123
Lion described by Buffon, 123—125
Lion described by Burchell, 125
Lion, disappearance of, 130
Lion and Hottentots, 132, 133—136
Lion-hunting Feats, 128
Lion, "King of the Forest," 126

INDEX. 351

Lion, Longevity of, 137
Lion, Maneless, 133—135
Lion, Niebuhr on, 131
Lion in the Nineveh Sculptures, 139, 140
Lions, the Drudhoe, 144
Lions, Popular Errors respecting, 123
Lion, Prickle or Claw in the Tail, 137—139
Lion, Roar of, 136
Lions in the Tower of London, 140
"Lion Tree" in the Mantatee Country, 127
Lion Stories of the Shows, 142
Lion-Talk, 123
Lioness and her Young, 135

MERMAID of 1822, 43—47
Mermaid in Berbice, 39
Mermaid in the Bosphorus, 47
Mermaid and Dugong, 41
Mermaids, Evidences of, 36
Mermaid at Exmouth, 40
Mermaid, Leyden's Ballad, 35
Mermaid and Manatee, 42
Mermaid at Milford Haven, 37
Mermaid, Japanese, 44
Mermaid, Scottish, 36, 38
Mermaids and Sirens, 33
Mermaid's Song, Haydn's, 34
Mermaids, Stories of, 33
Mermaid, Structure of, 43
Mermaids in Suffolk, 48
Mole, its Economy controverted, 62
Mole, the Ettrick Shepherd on, 71
Mole, Le Court on, 62, 65
Mole and Fairy Rings, 64
Mole and Farming, 70
Mole, Feeling of, 64
Mole at Home, 62
Mole, its Hunting-ground, 67
Moles, Loves of the, 68
Mole, structure of the, 63
Mole, St. Hilaire on, 69
Mole, Shrew, of North America, 70
Mole, Voracity of, 68
Montezuma's Zoological Gardens, 13
Musical Lizard, 303; Climbing Walls, 303, 304; Formosa Isle, 303; Gecko ennobled, 306

ORNITHOLOGICAL SOCIETY, 15
Ortolan described, 172, 173
Ortolans, how fattened, 174
Ortolan, Mr. Gould on, 174, 175
Owls, 221: Abyssinian Owl, 230; Barn Owl, 226; Biachaco, or Coquimbo, 224; Boobook Owl, 228; Cats and Owls, 230; Fraser's Eagle Owl, from Fernando Po, 229; Food of Owls, 226; Javanese Owl, 228; Snowy Owl, 227; Tricks by Night, 224; Utility of, 225; Waterton on the Owl, 225

PELICANS and Cormorants, 195
Pelicans described by Gould, 195
Pelican in Japan, 197
Pelican Popular Error, 198, 199
Pelican Pouches, 198
Pelican Symbol, 200
"Pelican of the Wilderness," 197
Pholas, Life and Labours of, 341
Pholades, Charlesworth and Peach on, 347
Pholades, Harper on, 346
Pholades, Robertson on, 343

RHINOCEROS in England, 22;
African Rhinoceros in 1858, 27;
Ancient History, 23; Bruce and Sparmann, 27; Burchell's shooting, 30; Horn of the Rhinoceros, 31, 32; Indian Wild Ass, 24; One-horned and Two-horned, 23-26; Scripture, Rhinoceros of, 23; Speehman's Rhinoceros Shooting, 30; Tegetmeir describes the African Rhinoceros, 27; Tractability, 25; Varieties of Rhinoceros, 22; Zoological Society's Rhinoceros, 23, 29

SALE of Wild Animals, 20
Sentinel Birds, 183
Song of the Cicada, 329
Songs of Birds and Seasons of the Day, 219
St. James's Park Menagerie, 14
Stories of the Barnacle Goose, 331-335
Stories of Mermaids, 33
Surrey Zoological Gardens, 20

TALKING BIRDS, 205; Bittern and Night Raven, 207; Blue Jay, 206; Canaries, Talking, 210-212; Chinese Starling, 205; Crowned Crane, 206; Cuckoo, 209; Laughing Goose, 209; Nightingale, 209; Piping Crow, 205; Snipe, Neighing, 213; Trochilos and Crocodile, 216; Umbrella Bird, 206; Whidaw Bird, 205; Wild Swan, 209; Woodpecker at Constantinople, 215
Talk about Toucans, 179; Bills of Toucans, 180; Carnivorous propensity, 184; Economy of, 182; Food of, 183; Gould, Mr., his Grand Monograph, 180, 186; Owen, Professor, on the Mandibles, 185; Swainson, Mr., on Toucans, 185

Toucan Family, 179, 180; White Ants' Nests, 183; Toucanet, Gould's, 184
Toad and Frog Concerts, 327-328
Toads, Running, Dr. Husenbeth's, 323-327
Tower of London Menagerie, 14
Tree-climbing Crab, the, 288: Bernhard, Hermit, and Soldier Crab, 291; Climbing Perch. 288; Crab, Burrowing, 290; Crab Migration in Jamaica, 292; Fishing-frogs, 288; Glass Crabs, 301; Pill-making Crabs, 301; Purse Crab feeding on Cocoanuts, 296; Robber Crab, 292; Screwpines, Crab climbing, 295; Vaulted Crab of the Moluccas, 291

UNICORNS, ancient, 51
Unicorn and Antelope, 53
Unicorn in Central Africa, 58
Unicorn described by Ctesias, 49, 50
Unicorn, Cuvier on, 54
Unicorn, Is it Fabulous? 49
Unicorn, Klaproth on, 55
Unicorn in Kordofan, 53
Unicorn and its Horn, 53, 59
Unicorn, modern, 50
Unicorn, Ogilby on, 51
Unicorn, Rev. J. Campbell on, 57
Unicorn in the Royal Arms, 60

WEATHERWISE ANIMALS, 231: Ants, Asses, 241; Darwin's Signs of Rain, 248; Frogs and Snails, 237-240; List of Animals, 241-247; Mole, 240; Mother Carey's Chickens and Goose, 234; Redbreast, 236; Seagulls, 232; Signs of Rain, 232; Stormy Petrels, 233; Shepherd of Banbury, 249; Toucans, 237; Weatherproof Birds' Nests, 247; Wild Geese and Ducks, 235
Wild Animals, Cost of, 19
Wild Beast Shows, 21

ZOOLOGICAL GARDENS, Origin of, 12
Zoological Society of London, 16
Zoology, Curiosities of, 11